Springer Theses

Recognizing Outstanding Ph.D. Research

Aims and Scope

The series "Springer Theses" brings together a selection of the very best Ph.D. theses from around the world and across the physical sciences. Nominated and endorsed by two recognized specialists, each published volume has been selected for its scientific excellence and the high impact of its contents for the pertinent field of research. For greater accessibility to non-specialists, the published versions include an extended introduction, as well as a foreword by the student's supervisor explaining the special relevance of the work for the field. As a whole, the series will provide a valuable resource both for newcomers to the research fields described, and for other scientists seeking detailed background information on special questions. Finally, it provides an accredited documentation of the valuable contributions made by today's younger generation of scientists.

Theses may be nominated for publication in this series by heads of department at internationally leading universities or institutes and should fulfill all of the following criteria

- They must be written in good English.
- The topic should fall within the confines of Chemistry, Physics, Earth Sciences, Engineering and related interdisciplinary fields such as Materials, Nanoscience, Chemical Engineering, Complex Systems and Biophysics.
- The work reported in the thesis must represent a significant scientific advance.
- If the thesis includes previously published material, permission to reproduce this must be gained from the respective copyright holder (a maximum 30% of the thesis should be a verbatim reproduction from the author's previous publications).
- They must have been examined and passed during the 12 months prior to nomination.
- Each thesis should include a foreword by the supervisor outlining the significance of its content.
- The theses should have a clearly defined structure including an introduction accessible to new Ph.D. students and scientists not expert in the relevant field.

Indexed by zbMATH.

Oscar Miguel-Escrig

Event-Based PID Controllers with Fixed Threshold Sampling Strategies

Doctoral Thesis accepted by
Universitat Jaume I, Castelló de la Plana, Spain

Springer

Author
Dr. Oscar Miguel-Escrig
Department of Industrial Systems
Engineering and Design
Universitat Jaume I
Castelló de la Plana, Spain

Supervisor
Prof. Julio-Ariel Romero-Pérez
Department of Industrial Systems
Engineering and Design
Universitat Jaume I
Castelló de la Plana, Spain

ISSN 2190-5053 ISSN 2190-5061 (electronic)
Springer Theses
ISBN 978-3-031-62504-6 ISBN 978-3-031-62505-3 (eBook)
https://doi.org/10.1007/978-3-031-62505-3

This Springer imprint is published by the registered company Springer Nature Switzerland AG
The registered company address is: Gewerbestrasse 11, 6330 Cham, Switzerland

To my parents

Supervisor's Foreword

It is with great pleasure that I present this thesis, which explores the robustness and tuning of event-based Proportional-Integral-Derivative (PID) controllers in networked control systems. This work represents a culmination of diligent research and analysis conducted by Oscar Miguel-Escrig, shedding light on critical aspects of controller design in the context of event-driven control strategies.

The thesis delves into the intricate dynamics of discrete PID controllers operating under various sampling strategies, including Symmetric-Send-On-Delta (SSOD) and Regular Quantization with Hysteresis (RQH). Through meticulous investigation and innovative methodologies, Oscar Miguel-Escrig uncovers novel insights into the robustness requirements and tuning procedures necessary to mitigate limit cycle oscillations and optimize controller performance in networked environments.

One of the standout contributions of this thesis is the development of robustness measures tailored specifically for event-based controllers, like the Tsypkin margin (M_T), which provides a quantitative assessment of a controller's ability to avoid limit cycle oscillations induced by the sampling strategy. By introducing such measures and leveraging advanced techniques like the Sampled Describing Function, Oscar Miguel-Escrig offers a comprehensive framework for evaluating and enhancing the robustness of event-based control systems.

Furthermore, this thesis bridges the gap between theoretical analysis and practical implementation by providing concrete guidelines and tuning procedures for event-based controllers. The proposed methodologies are supported by extensive simulation studies and empirical validation, demonstrating their effectiveness in real-world scenarios.

I commend Oscar Miguel-Escrig for their exceptional dedication and scholarly rigor in conducting this research. I am confident that this thesis will not only contribute significantly to the academic literature but also serve as a valuable resource for practitioners and researchers in the field of control systems engineering.

Castelló de la Plana, Spain
March 2024

Dr. Julio-Ariel Romero-Pérez

Foreword from CEA, the Automatic Control Spanish Association

It is a great pleasure for us to introduce here the work of Dr. Oscar Miguel-Escrig, awarded with the best *Ph.D. Thesis in Control Engineering* during the 2021–2023 edition of the award call organized by the control engineering group of *Comité Español de Automática (CEA), the Spanish Committee of Automatic Control.*

This two-year award is aimed at recognizing outstanding Ph.D. research carried out within the Control Engineering field. At least one of the supervisors must be partner of the Spanish Committee of Automatic Control and member of the Control Engineering Group. The jury is composed by three well-known doctors in the control engineering field; two of them are partners of the CEA, while the third one is a foreign professor.

This edition counted submissions from Ph.D. students examined between September 2021 and July 2023. Notably, the scientific production of the candidates consisted of a total of 58 publications in international indexed journals, where 33 of them are ranked in first quartile journals. This shows the high scientific quality of the submitted Ph.Ds. All the thesis presented:

- have been under the international doctorate mention (the candidate must have at least one international research stay out of Spain, write part of the thesis in English, and have foreign members in the Ph.D. jury).
- have been mostly oriented to academic research than to industrial applications.
- were presented at international conferences. Among them, IFAC and IEEE are considered as preferred conferences.

As a result, Oscar Miguel-Escrig's Ph.D. thesis was selected as the best one among an excellent group of candidates. His Ph.D. thesis deserves the label "the best of the best", Springer Theses' main motto. Not only his nomination as winner of the Spanish control engineering context has been our great pleasure, but also to recommend his thesis for publication within the Springer Theses collection. On behalf of CEA we wish Dr. Miguel to continue his outstanding scientific career, keeping his genuine enthusiasm. We also hope that his work may be of inspiration for other students working in the control engineering field.

Prof. Dr. José Manuel Andújar
President of CEA

Dr. Manuel G. Ortega
Dr. Javier Rico
Coordinators of the Control
Engineering Group
Zaragoza, Spain

Parts of this thesis have been published in the following articles

Journal Papers

1. Oscar Miguel-Escrig and Julio-Ariel Romero-Pérez. Regular quantisation with hysteresis: a new sampling strategy for event-based PID control systems. IET Control Theory & Applications, 14(15):2163–2175, 2020.
2. Oscar Miguel-Escrig, Julio-Ariel Romero-Pérez, and Roberto Sanchis-Llopis. Tuning PID controllers with symmetric send-on-delta sampling strategy. Journal of the Franklin Institute, 357(2):832–862, 2020.
3. Oscar Miguel-Escrig and Julio-Ariel Romero-Pérez. Tuning procedure for event-based PI controllers under regular quantization with hysteresis. Journal of the Franklin Institute, 358(9):4643–4670, 2021.
4. Oscar Miguel-Escrig, Julio-Ariel Romero-Pérez, José Sánchez-Moreno, and Sebastian Dormido. Characterization of limit cycle oscillations induced by fixed threshold samplers. IEEE Access, 10:62581–62596, 2022.
5. Oscar Miguel-Escrig and Julio-Ariel Romero-Pérez. Event-based discrete PI controllers robustness analysis through sampled describing function technique. International Journal of Control, 97(1):86–100, 2024.

Conference Papers

1. Oscar Miguel Escrig, Romero-Pérez Julio Ariel, and Esteban Querol-Dolz. Implementación y evaluación de controladores basados en eventos en la norma IEC-61499. Actas de las XXXVIII Jornadas de Automática, 2017.
2. Oscar Miguel-Escrig and Julio-Ariel Romero-Pérez. Generación extrínseca de eventos para el control de sistemas continuos. Actas de las XXXIX Jornadas de Automática, 2017.
3. Oscar Miguel-Escrig, Julio-Ariel Romero-Pérez, and Roberto Sanchis-Llopis New robustness measure for a kind of event-based PID. IFACPapersOnLine 51(4): 781–786, 2018.
4. Oscar Miguel-Escrig and Julio-Ariel Romero-Pérez. Implementation and evaluation of event-based PID in the iec-61499 standard. In 2018 IEEE 23rd International Conference on Emerging Technologies and Factory Automation (ETFA), volume 1, pages 574–580. IEEE, 2018.
5. Oscar Miguel-Escrig and Julio Ariel Romero Pérez. Mejora de la estrategia de muestreo en bucles de control PID basados en eventos. Actas de las XXXIX Jornadas de Automática, 2018.

6. Oscar Miguel-Escrig and Julio-Ariel Romero-Pérez. Event-based controllers with external event generation implementation issues and computational cost study. In 2019 IEEE 58th Conference on Decision and Control (CDC), pages 722–727. IEEE, 2019.
7. Oscar Miguel-Escrig and Julio-Ariel Romero-Pérez. Robustness of event based discrete PI controllers. A sampled describing function approach. IFAC-PapersOnLine, 53(2): 2796–2801, 2020.
8. Oscar Miguel-Escrig and Julio-Ariel Romero-Pérez. Implementation and experimental evaluation of SSOD sampling strategy for ebc. In 2021 26th IEEE International Conference on Emerging Technologies and Factory Automation (ETFA), pages 1–8. IEEE, 2021.

Acknowledgements and Declarations

First of all, I want to thank my parents for always being there, for letting me dream high while having my feet on the ground. I would also like to thank my *wela* Rosa for making me smile. And also, to those who have left, I am taking you well inside.

Secondly, to Raquel, for all those things you do and for which I don't thank you enough. I love you.

Thirdly, to all the colleagues with whom I work side by side and share both the sweet and the bitter moments. And obviously to Julio, a great person and a friend, who happens to be my thesis supervisor.

Finally, I would also like to thank my lifetime friends, those who, no matter what happens or how long it has been since we have seen each other, will always be there. We keep walking together.

Competing Interests

This study was primarily funded by Conselleria d'Educació, Investigació, Cultura i Esport grant number ACIF/2018/244.

This study also received funding from various projects:

"Adaptación de algoritmos de control PID al contexto de la Industria 4.0" from Universitat Jaume I (Ref. UJI-B2018-39)

"Desarrollo de algoritmos de control para sistemas distribuidos implementables mediante estándares industriales" from Universitat Jaume I (Ref. P11B2015-42)

"Estrategias de estimación y control para la minimización del coste energético en procesos con perturbaciones variantes y sujetos a periodos tarifarios" from Ministerio de Ciencia e Innovación (Ref. TEC2015-69155-R).

Contents

Acronyms

Abbreviations

ADC	Analog to Digital Converter
DF	Describing Function
DIDF	Dual Input Describing Function
EBC	Event-Based Control
EBPID	Event-Based PID
ECC	Execution Control Chart
FB	Function Block
FO	Fractional Order
FOPTD	First Order Plus Time Delay
FTS	Fixed Threshold Sampler
I$\bar{\text{E}}$	Integral of the sampled Error
IA$\bar{\text{E}}$	Integral of the Absolute value of the sampled Error
IAE	Integral of the Absolute value of the Error
IE	Integral Error
PID	Proportional-Integral-Derivative controller
PLC	Programmable Logic Computer
PWM	Pulse Width Modulated signal
RQ	Regular Quantization
RQH	Regular Quantization with Hysteresis
SOD	Send-On-Delta
SSOD	Symmetric-Send-On-Delta
ZOH	Zero Order Hold

Symbols

$\mathbf{e_{ss}}$	Steady-state error
$\delta_{\mathbf{u}}$	Control action change due to a variation δ
$\mathbf{G_{ol}}$	Open-loop transfer function
$\mathcal{N}, \mathcal{N}_{\mathbf{A}}$	Describing function for sustained oscillations
$\mathcal{N}_{\mathbf{B}}$	Describing function for sustained bias
$\mathcal{N}'$	Sampling time normalized describing function
$\Re\{\cdot\}$	Real part of complex number
$\Im\{\cdot\}$	Imaginary part of complex number
$\gamma_{\mathbf{h}/\delta}$	Gain margin to non-linearity
$\Phi_{\mathbf{h}/\delta}$	Phase margin to non-linearity
$\lvert \cdot \rvert$	Modulus of complex number
$\arg\{\cdot\}$	Phase of complex number
$\gamma_{\mathbf{cg}}$	Classical gain margin
$\Phi_{\mathbf{cp}}$	Classical phase margin
$\mathbf{t_s}$	Settling time
$\mathbf{M_T}$	Tsypkin margin
$\mathbf{B_T}(\omega_{\mathbf{o}})$	Tsypkin branch corresponding to frequency ω_o
$\bar{\mathbf{s}}$	Dimensionless Laplace variable
$\overline{\mathbf{K_p}}$	Dimensionless controller gain
$\overline{\mathbf{T_i}}$	Dimensionless integral time
$\overline{\mathbf{T_d}}$	Dimensionless derivative time
$\bar{\mathbf{G}}(\bar{\mathbf{s}})$	Dimensionless transfer function
$\mathbf{C}(\infty)$	Noise amplification
$\bar{\mathbf{C}}(\infty)$	Dimensionless noise amplification
$\mathbf{R_m}$	Tsypkin robustness against m-leveled oscillations
$\mathbf{C(z)}$	Z-transform of controller
$\mathbf{C^*(s)}$	Starred transform of controller
$\mathbf{D_{ch}}$	Distance to convex hull

List of Figures

List of Tables

Chapter 1
Introduction

1.1 Background

Over the past few decades, there has been a growing utilization of Event-Based Control (EBC) for continuous systems in industrial applications. In contrast to the conventional time-triggered controllers employed in industry, which execute control algorithms at fixed intervals, Event-Based Controllers initiate their control algorithms solely upon the occurrence of asynchronous events that indicate noteworthy changes in the system's state. An illustrative instance of such an event reception strategy could be a sampler triggering an event whenever the error signal (difference between set-point or reference and controlled output) surpasses specific thresholds.

EBC pursues two paramount objectives:

1. Reducing the amount of necessary data to perform the closed-loop control.
2. Diminishing the average computational cost required by control algorithms.

Owing to these attributes, EBC algorithms have found primary application in distributed control systems, aiming to minimize data traffic across industrial communication networks. This is particularly relevant in the interaction between sensor nodes, controllers, and actuators. Various control strategies, including the PID controller—widely employed in industry [1]—have been adjusted to suit EBC, counting with multiple versions and tuning methods specifically tailored for this context.

One of the initial advancements in the creation of Event-Based PID controllers was presented by Årzén in [2]. This Event-Based Controller effectively reduced CPU utilization in computer-controlled EBC systems without significantly impacting the behavior of the control loop. Some of the limitations induced by early stage implementations of this kind of algorithms were spotted in this contribution, notably, the errors that arise in the calculation of integral and derivative terms as the time between samples increases. Subsequent research efforts have been dedicated to address the issues identified by Årzén, particularly those associated with errors in the computation of the integral term. Noteworthy contributions in this regard come from Durand [3, 4] and Vasyutynskyy [5, 6].

O. Miguel-Escrig, *Event-Based PID Controllers with Fixed Threshold Sampling Strategies*, Springer Theses, https://doi.org/10.1007/978-3-031-62505-3_1

Concerning the event generation strategy applied in EBC, the predominant approach, particularly when employing a PID controller, involves sending new data when the sampled signal surpasses specific value thresholds, denoted as δ. This event generation policy is referred to as Send-On-Delta (SOD). The effectiveness of SOD has been assessed in terms of control loop performance and reduction of communication data flow in studies such as [7, 8].

An adaptation of the SOD sampling method for PI control was initially suggested in [9]. Unlike in SOD sampling, this variant involves quantifying the signal in multiples of the minimum threshold δ. Consequently, this adjustment ensures that the relationship between the input and output of the event generator is symmetric with regard to its origin. This variant is called Symmetric-Send-On-Delta (SSOD), and has been explored in preliminary studies in loops with a PI controller [9, 10]. Additionally, a tuning procedure for PI controllers under SSOD sampling has been proposed for first order plus time delay (FOPTD) systems in [11].

Recent findings have been disclosed in [12–15]. In [12], the authors introduce a tuning guideline for PID controllers based on SSOD sampling. This tuning approach relies on the utilization of phase and gain margins, parameters commonly employed in classical methods for controller design, facilitating its practical application. Similarly, in [14], the tuning methodology is refined, ensuring optimal controller performance concerning perturbation response by minimizing the integral error. Simultaneously, it guarantees robustness against limit cycle oscillations induced by the SSOD sampler.

Conversely, in [13], the impact of incorporating different sampling schemes into the PID control loop was investigated. Specifically, the examined sampling strategies were SSOD and Regular Quantification (RQ). Employing the Describing Function (DF) technique as an analytical tool, the study derived the conditions that a system must satisfy to avoid limit cycles induced by these sampling techniques. These conditions were expressed in both cases in terms of gain and phase margins. Furthermore, [15] conducted a comparative analysis of the robustness achieved with RQ and SSOD. Additionally, the advantages and disadvantages of each sampling strategy were elucidated.

Alternative approaches to Event-Based PID controllers are feasible; for example, in [16], a controller with adaptive thresholds was suggested. This adaptable sampling strategy achieved a trade-off between the quantity of generated events and the magnitude of perturbations.

1.1.1 Implementation of EBC According to Standard IEC 61499

In the industrial automation programming context, various approaches exist for developing applications, most of them align with the principles outlined in the IEC 61131 standard [17]. However, this standard has certain limitations, with the primary concern being the challenge of distributing applications across multiple devices. To address this issue, the IEC 61499 standard [18] was introduced, aiming to simplify the design and implementation of distributed and reconfigurable applications. A notable distinction between these standards lies in their execution models. While IEC 61131 relies on the periodic execution of application algorithms, IEC 61499 adopts an event-driven execution model, supported by the management and handling of events.

While there is a substantial body of research focused on the application of IEC 61499 in discrete Event-Based Control systems, the same cannot be asserted for its utilization in continuous control systems, despite its significance in the industrial context.

In the context of this research line, there is notable work, like [19], which stands out as one of the initial efforts emphasizing the necessity of instilling a deterministic behavior, in terms of execution times, to IEC 61499 to establish reliable control applications. In [20], it was conducted an experimental study on the suitability of the standard for controlling a robotic arm using a PID controller, demonstrating its viability in such applications. Other works, like [21], underscore some essential considerations for implementing controllers for continuous systems within the framework of IEC 61499. In [22], the IEC 61499 standard was employed to implement distributed control for a flexible manufacturing system. With regard to the event management, [23] presented a study highlighting the importance of a proper event management system for controlling continuous systems.

In spite of the endeavors in this research direction, the majority of studies involving IEC 61499 for continuous control predominantly focus on time-triggered controllers. By adopting these control algorithms, the characteristic event-driven execution model of IEC 61499 programming is overlooked. This aspect of the standard implies that EBC might be a more fitting algorithm for controlling continuous systems.

1.2 Challenging Problems

Recent research in the field of Event-Based PID controllers has yielded numerous compelling results. Nevertheless, several crucial questions remain unanswered. The primary aim of this thesis is to address some of these challenging issues.

Development of Specific Analysis and Tuning Methods
The literature has introduced numerous control schemes incorporating event-based PID controllers, yet insufficient attention has been given to the tuning methods for designing these controllers. Actually, the tuning methods employed in these works are derived from continuous control loop schemes or, at best, from sampled or hybrid control schemes. The extension of these methods to the actual event-based scenario often relies on empirical ideas rather than a thorough examination of stability conditions, robustness margins, and system performance. Additionally, Event-Based Controllers are commonly implemented in microprocessor-based devices like PLCs in industrial settings or cards such as Raspberry Pi and BeagleBone Black.

Controllers on these devices are typically executed as periodic routines, with specific guidelines for choosing sampling and control periods, which are critical parameters. However, in the most analogous version of this implementation, i.e., periodic triggered Event-Based Control, these guidelines are not applicable. Therefore, dedicated analysis tools are needed to assess the performance and robustness of a given system, along with tuning rules based on these analyses to ensure optimal performance and robustness of the controlled system.

Study of New Sampling Schemes
Event-Based PID controllers have predominantly employed sampling schemes based on threshold crossings, with SOD and some of its variations, like SSOD. However, certain studies suggest that alternative sampling strategies may offer advantages in terms of event generation and robustness, depending on the specific application, without compromising loop performance. The adoption of these new sampling strategies will necessitate the development of dedicated analysis methodologies and tuning methods for controllers, tailored to the characteristics of each sampling strategy.

Implementation of Event-Based PID Controllers in the Standard IEC 61499
Hitherto, systematic methods for implementing Event-Based Control applications within the framework of the IEC 61499 standard are lacking. Additionally, controllers employed for regulating continuous processes according to this standard typically function as discrete controllers with periodic execution, missing out on the potential benefits offered by the event-based execution capabilities of IEC 61499. Thus, it appears that there is a synergistic relationship between Event-Based Control systems and the IEC 61499 standard. Event-Based Control provides suitable algorithms and control methodologies for the distributed control of continuous processes to the standard, while the standard streamlines the implementation, configuration, verification, and maintenance of the networked control system. It is worth noting that these control algorithms can enhance computation time without significantly impacting the performance of the control loop.

1.3 Thesis Outline

This thesis aims to provide techniques and methodologies to study the robustness of loops in which Event-Based PID controllers are involved. Various sampling methodologies relying on fixed-threshold quantization, responsible for event generation, are examined. Firstly, the characterization of robustness is addressed, and, secondly, several design methodologies are introduced, considering both robustness and control loop performance. Hence, the chapters' outcomes revolve around these concerns, each focusing on different aspects or employing distinct approaches.

The present theses is sectioned in two parts. Part I presents the analysis conducted using several techniques, ultimately leading to the characterization of the robustness through several measures. Chapters 2–5 are dedicated to the study of SSOD, RQH and in general to FTS sampling strategies and their effect on the loop performance and robustness under different theoretical frameworks.

Part II presents different tuning methodologies for Event-Based Control and their implementation. The presented tuning rules are grounded on the robustness measures defined in Part I, and also take into consideration classical robustness margins to better assess performance. For the implementation, standard IEC 61499 is used, which aims to standardize the implementation of distributed control systems and in particular, the Event-Based Control framework under study. Chapters 6–8 are dedicated to these issues.

Each chapter has been conceived to allow its lecture independently of the others. Presented contents partly derive from material in press or published papers and, therefore, some repetitions may appear on the preliminaries of each chapter. A summary of the content in each chapter can be freely consulted through their abstract.

References

1. Samad T (2017) A survey on industry impact and challenges thereof. IEEE Control Syst 37(1):17–18
2. Årzén K-E (1999) A simple event-based PID controller. In: Proceedings 14th IFAC world congress vol 18, pp 423–428
3. Durand S, Marchand N (2009) An event-based PID controller with low computational cost. In: 8th international conference on sampling theory and applications (SampTA'09), pp Special–session
4. Durand S, Marchand N (2009) Further results on event-based PID controller. In: Control conference (ECC), 2009 European. IEEE, pp 1979–1984
5. Vasyutynskyy V, Kabitzsch K (2009) Time constraints in PID controls with send-on-delta. IFAC Proc Vol 42(3):48–55
6. Vasyutynskyy V, Kabitzsch K (2010) A comparative study of PID control algorithms adapted to send-on-delta sampling. In: 2010 IEEE international symposium on industrial electronics (ISIE). IEEE, pp 3373–3379
7. Dormido S, Sánchez J, Kofman E (2008) Muestreo, control y comunicacián basados en eventos. Revista Iberoamericana de Automática e Informática Industrial RIAI 5(1):5–26
8. Ploennigs J, Vasyutynskyy V, Kabitzsch K (2010) Comparative study of energy-efficient sampling approaches for wireless control networks. IEEE Trans Ind Inf 6(3):416–424
9. Beschi M, Dormido S, Sánchez J, Visioli A (2012) Characterization of symmetric send-on-delta PI controllers. J Process Control 22(10):1930–1945
10. Chacón J, Sánchez J, Visioli A, Yebra L, Dormido S (2013) Characterization of limit cycles for self-regulating and integral processes with PI control and send-on-delta sampling. J Process Control 23(6):826–838
11. Beschi M, Dormido S, Sánchez J, Visioli A (2014) Tuning of symmetric send-on-delta proportional-integral controllers. IET Control Theory Appl 8:248–259(11)
12. Romero-Pérez J-A, Sanchis-Llopis R, Peñarrocha-Alós I (2014) A simple rule for tuning event-based PID controllers with symmetric send-on-delta sampling strategy. In: Proceedings of the 2014 IEEE emerging technology and factory automation (ETFA), pp 1–8, Sept 2014
13. Romero-Pérez J-A, Sanchis-Llopis R, Arrebola E (2015) Experimental study of event based PID controllers with different sampling strategies. Application to brushless DC motor networked control system. In: 2015 XXV international conference on information, communication and automation technologies (ICAT), pp 1–6, Oct 2015

14. Romero-Pérez J-A, Sanchis-Llopis R (2016) A new method for tuning PI controllers with symmetric send-on-delta sampling strategy. ISA Trans 64:161–173
15. Romero-Pérez J-A, Sanchis-Llopis R (2017) Tuning and robustness analysis of event-based PID controllers under different event generation strategies. Int J Control 1–38
16. Romero-Pérez J-A, Pascual NJ, Peñarrocha-Alós I, Sanchis-Llopis R (2012) Event-based PI controller with adaptive thresholds. In: 2012 4th international congress on Ultra modern telecommunications and control systems and workshops (ICUMT). IEEE, pp 219–226
17. John K-H, Tiegelkamp M (2010) IEC 61131-3: programming industrial automation systems: concepts and programming languages, requirements for programming systems, decision-making aids. Springer Science & Business Media
18. Zoitl A, Lewis R (2014) Modelling control systems using IEC 61499, vol 95. IET
19. Thomas Strasser, Franz Auinger, and Alois Zoitl. Development, implementation and use of an IEC 61499 function block library for embedded closed loop control. In *Industrial Informatics, 2004. INDIN'04. 2004 2nd IEEE International Conference on*, pages 594–599. IEEE, 2004
20. Doukas GS, Thramboulidis KC, Koveos YC (2006) Using the function block model for robotic arm motion control. In: 14th mediterranean conference on control and automation, 2006. MED'06. IEEE, pp 1–6
21. Hametner R, Schitter G, Voigt A, Zoitl A (2013) Implementation guidelines for closed loop control algorithms on PLCs. In: 2013 IEEE international conference on industrial technology (ICIT). IEEE, pp 1297–1302
22. Catalán-Cantero C, Modelos y plataforma IEC 61499 adaptados al control distribuido de máquinas herramienta en sistemas de fabricación ágil
23. Querol-Dolz E, Romero-Pérez J-A, Serrano-Mira J, Sanchis-Llopis R (2016) Evaluation of closed loop control applications using different event management strategies under IEC 61499. In: 2016 second international conference on event-based control, communication, and signal processing (EBCCSP). IEEE, pp 1–8

Part I
Analysis of Event-Based Systems

Chapter 2
RQH Sampling Strategy Analysis with DF Method

2.1 Introduction

In recent years, numerous publications have emerged focusing on Event-Based Control (EBC) for continuous systems, as highlighted in [1]. EBC offers the advantage of optimizing data flow in networked control systems, thereby reducing packet losses and minimizing delays introduced by communication infrastructures. This efficiency stems from the practice of sending new data only when significant changes are detected in the system's state, contrasting with the periodic approach of time-driven controllers. Consequently, EBC stands out as a promising control methodology in networked control systems, a recognition underscored in the context of modern factory automation, as mentioned in [2]. For an in-depth and updated examination of the primary contributions to EBC over the last two decades, the reader can consult the comprehensive study provided in [3].

Among the control strategies tailored to the EBC paradigm, particular attention has been drawn to the PID algorithm. It is indisputable that PID controllers are present ubiquitously in most industrial control applications, with statistics provided in [4] indicating that over 95% of controllers fall under this category. Echoing this prevalence, a survey conducted among members of the industrial committee within the International Federation of Automatic Control, as reported in [5], underscores the dominant position of PID algorithms compared to other advanced control strategies such as Model Predictive Control (MPC). The pivotal role of PID in the context of Industry 4.0 has recently been emphasized in [6], recognizing the need for incorporating new features to align this well-established control technology with the distinctive characteristics of this emerging paradigm. Industry 4.0 is characterized by high connectivity between devices through wired and wireless communication networks. This, coupled with the established dominance of PID controllers in industry, has spurred extensive research on Event-Based PID controllers in recent decades.

To the best of the author's knowledge, one of the earliest works on Event-Based PID (EBPID) was introduced in [7]. This paper addressed issues related to errors

O. Miguel-Escrig, *Event-Based PID Controllers with Fixed Threshold Sampling Strategies*, Springer Theses, https://doi.org/10.1007/978-3-031-62505-3_2

in calculating the integral and derivative terms when irregular sampling leads to an increase in the time between samples. Subsequently, this problem received thorough treatment in later works by Durand [8, 9] and Vasyutynskyy [10, 11]. While the primary objective in [7] was to employ EBPID for reducing CPU usage in embedded control systems without significantly affecting closed-loop performance, the foundational ideas presented in that paper were later expanded to encompass the application of EBPID in networked control systems.

As demonstrated early on [12], the choice of sampling law significantly influences the performance and behavior of sampled control systems. In the context of Event-Based Control systems, the employed sampling strategies play a crucial role as they are responsible for generating events for the execution of the controller's algorithm. Among these strategies, those based in signal quantification have earned increased significance due to their straightforward implementation. A prominent example is Send-On-Delta (SOD) sampling, where the transmission of the signal value occurs solely when it crosses predefined levels or thresholds of magnitude δ. The effectiveness of this strategy has undergone extensive testing, evaluating its impact on control performance and communication reduction [13, 14].

Based on SOD ideas, a sampling strategy termed Symmetric-Send-On-Delta (SSOD) was introduced in [15]. SSOD is characterized by incorporating hysteresis of the same value as the thresholds δ into the sampler and fixed commutation levels. Over the past years, various studies have delved into SSOD-based PI controllers, encompassing both tuning methodologies and practical applications. In [16], the tuning of SSOD-PI controllers for FOPTD systems was addressed, with rules designed to minimize the 1% settling time of the closed-loop response. An application of SSOD loops in controlling the inside air temperature of greenhouse production processes was presented in [17]. Additionally, tuning methods for PI controllers with SSOD sampling were developed in [18, 19], employing new robustness margins derived from the Describing Function (DF) technique to avoid limit cycles, and connecting with classical concepts of phase and gain margins. In [20] a unified design of SSOD-based PID and Smith predictor for self-regulating and integral processes was explored. A novel system identification procedure, based on oscillations induced by the SSOD sampling strategy, was proposed in [21]. More recently, [22] introduced a procedure for tuning not only PI but also PID controllers with SSOD sampling for FOPTD systems. This method relies on defining a new robustness measure, known as the Tsypkin margin and presented in [23], which overcomes the limitations associated with the DF approach. Collectively, these works underscore the growing interest in SSOD-based controllers over the past decade.

As an alternative to SSOD, another sampling strategy is the Regular Quantification (RQ), wherein new data is transmitted whenever the sampled signal's value is a multiple of the quantification threshold δ. A comparative study between SSOD and the RQ strategy has been detailed in [24]. The absence of hysteresis in the RQ sampler may lead to bursts of events triggered by measurement noise, constituting the primary drawback of RQ compared to SSOD. Conversely, in RQ loops, there is greater robustness against the emergence of limit cycles than in SSOD loops, even with similar loop performances. These findings suggest that sampling strategies lying between RQ and SSOD could potentially offer superior overall behavior, as they may mitigate event generation due to noise without imposing as restrictive robustness requirements as those in SSOD samplers.

The aforementioned research in the field of EBPID indicates that, in general, conventional PID controllers operating under event-based sampling strategies maintain a favorable trade-off between performance and robustness while significantly reducing the required data for control. As a consequence, the transmission through digital networks in networked systems is also markedly reduced. The results presented in this chapter further explore this approach by introducing new sampling strategies for Event-Based PI controllers, aiming to address the limitations of SSOD and RQ samplers.

In this chapter a new sampling law is proposed, which can be considered as a generalization of SSOD and RQ. The approach involves independently selecting the hysteresis and quantification threshold, allowing for the separation of immunity to noise from the controller's responsiveness to significant changes in the system. Additionally, it is demonstrated that the new proposal can significantly decrease the number of generated events compared to the SSOD strategy. The new sampler is thoroughly examined in terms of robustness, with the development of quantitative measures to prevent the occurrence of limit cycle oscillations.

Given the sensitivity of the derivative action to measurement noise in PID controllers, the PI form of the controller prevails over the PID form [25]. This was illustrated by a study in [26], where over 94% of controllers in power plants in Guangdong province, China, are identified as PI controllers. Due to this prevalence, the robustness measures proposed in this chapter are utilized to evaluate the applicability of well-known tuning methods commonly used for designing continuous PI controllers when applied to RQH-PI loops. Specifically, the Ziegler-Nichols [27], Cohen-Coon [28], and AMIGO [29] methods are assessed. These methods are easily applicable in industry as they allow for straightforward calculation of controller parameters using simple equations and data obtained from step response or relay feedback experiments. Previous studies by the author have confirmed that under certain cases the AMIGO method provides proper tuning for this control scheme when SSOD sampling is employed [22]. Among other findings, this chapter's study demonstrates the effectiveness, or range of effectiveness, of each of these methods in preventing limit cycle oscillations when utilizing PI with RQH sampling.

2.2 Problem Statement

Consider the scheme of a networked control system presented in Fig. 2.1, where $C(s)$ and $G_p(s)$ are the controller and process transfer functions respectively, the EG block is a given event generator, the ZOH block is a zero-order hold and $\exp(-t_d s)$ models the network's delay. Besides, y_r is the reference signal, y the controlled output and p the disturbance input. It is assumed that the controller is located near to the actuator and, also, that the EG sends the measured signal e^* of the error e through the network

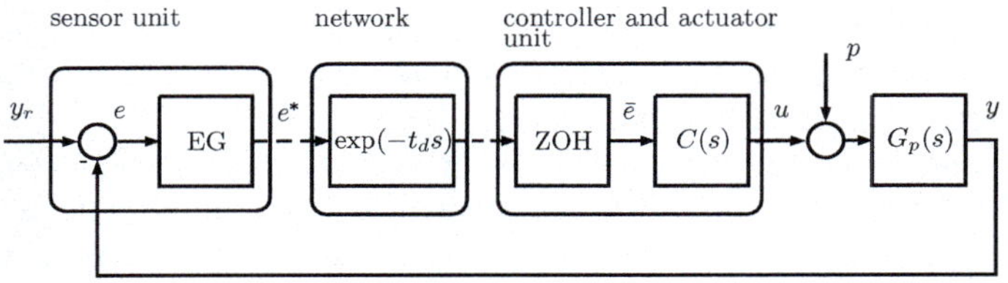

Fig. 2.1 Control loop scheme for event-based PID controllers proposed in [15]. (Reprinted from IFAC-PapersOnLine, Vol 51, No. 4, Miguel-Escrig, O., Romero-Pérez, J. A., & Sanchis-Llopis, R., *New robustness measure for a kind of event-based PID*, 781–786, Copyright (2018), with permission from Elsevier)

and the ZOH block holds that value in $\bar{e}$ until new data arrive. This control scheme was firstly proposed in [15] where $C(s)$ was considered to be a PI controller and the EG block a SSOD sampler. Therefore, the authors named that architecture SSOD-PI. Generally speaking, any controller and event generator can be used, thus the general naming EG-$C(s)$ will be used.

As aforementioned, the main sampling strategies used for the block EG using fixed thresholds are the SSOD and RQ.

As pointed in [24], by choosing a SSOD sampling the extra events due to noise can be avoided as long as its amplitude is lower than the hysteresis, whose value is equal to the threshold δ. On the other hand, the RQ sampler is very sensitive to noise, owing to the lack of hysteresis, producing bursts of events when the signal takes values close to the commutation levels.

However, with regard to the robustness, the SSOD-$C(s)$ configuration is more likely to propitiate the apparition of limit cycles. This is due to the effect of the hysteresis, thus, to avoid this oscillations higher robustness on the controller is required than in the RQ-$C(s)$ case, which admits controllers with lower robustness requirements. These potentially harmful consequences of applying this kind of samplers must be reduced to improve the overall performance of the loop. To this end, a new sampling strategy is introduced in the following lines, whose properties are studied by defining new robustness measures against limit cycle oscillations induction.

Remark 2.1 Oscillatory responses cannot be fully avoided for processes with integrator using the control architecture in Fig. 2.1 with the samplers under study. The process will continuously integrate any non-null control action, resulting in the sensor unit issuing an event and therefore entering the oscillatory state. For that kind of processes, the reader is urged to consult the work presented in [20] addressing this issue. Therefore, processes with integrator are out of the scope of this chapter.

2.3 RQH Sampling Strategy

SSOD and RQ exhibit complementary behaviors regarding to their strengths and weaknesses. Thus, a novel sampling strategy, which is an intermediate case, that gets advantage of both is proposed. This strategy has been named *Regular Quantization with Hysteresis* (RQH), and its input-output relationship is defined by Eq. (2.1),

$$\bar{x}(t) = \begin{cases} (i+1)\delta \text{ if } x(t) \geq \left(i + \frac{1}{2} + \frac{h}{2\delta}\right)\delta \text{ and } \bar{x}(t^-) = i\delta, \; i \in \mathbb{Z} \\ (i-1)\delta \text{ if } x(t) \leq \left(i - \frac{1}{2} - \frac{h}{2\delta}\right)\delta \text{ and } \bar{x}(t^-) = i\delta \\ i\delta \quad \text{if } x(t) \in \left[\left(i - \frac{1}{2} - \frac{h}{2\delta}\right)\delta, \left(i + \frac{1}{2} + \frac{h}{2\delta}\right)\delta\right] \text{ and } \bar{x}(t^-) = i\delta \end{cases} \tag{2.1}$$

where $x(t)$ and $\bar{x}(t)$ are the input output signals respectively.

This sampler is defined by two parameters, the quantization level $\delta > 0$ and the hysteresis h, which can be freely selected in the range $0 \leq h \leq \delta$. Specific cases can be found by fixing $h = 0$, obtaining RQ sampler, or $h = \delta$, resulting in SSOD. Therefore, RQH results in a generalization of those cases.

For in-between values of h characteristics of both SSOD and RQ can be observed. In general, increasing h/δ tends to reduce the burst of generated events due to the noise. Including a sufficiently high hysteresis completely eliminates them. This hysteresis has to be chosen so that the noise magnitude lays within its thresholds. Thus, the first guideline is to select h, at least, slightly higher than the noise peak-to-peak amplitude.

These sampling strategies under study introduce a steady-state error on the system output of Fig. 2.1 since within a band around $e = 0$ the controller will receive a null sampled error $\bar{e}$, however, e takes a value between the commutation thresholds. This band should correspond to a not significant change on the process output y. Equation (2.2) relates the admissible steady-state error (e_{ss}) with RQH parameters.

$$e_{ss} = \frac{1}{2}(\delta + h) \tag{2.2}$$

This constitutes the second guideline to select RQH parameters since h has already been selected, and δ is chosen from this expression to fulfill a given steady-state error requirement.

2.3.1 Event Generation

Reducing the communication output is a main goal in the design of networked control systems. In the case of EBC, this goal translates to reducing the number of generated events. For the proposed RQH sampler, a magnitude change of C generates a number of events n_{ev} defined according to:

$$n_{ev} = \left\lfloor \frac{C - \frac{1}{2}(\delta + h)}{\delta} + 1 \right\rfloor . \tag{2.3}$$

Combining expressions (2.2) and (2.3) the relationship between n_{ev}, h/δ, and C results in:

$$n_{ev} = \left\lfloor \frac{C - e_{ss}}{2e_{ss}} \left(1 + \frac{h}{\delta}\right) + 1 \right\rfloor . \tag{2.4}$$

Expressing C as multiples of the steady-state error as $C = e_{ss}\rho$, then:

$$n_{ev} = \left\lfloor \frac{\rho - 1}{2} \left(1 + \frac{h}{\delta}\right) + 1 \right\rfloor . \tag{2.5}$$

This expression defines the number of events produced by the RQH sampling strategy against a change of ρ times the steady-state error. The ratio h/δ influences significantly the number of events for the same value of ρ. In particular, the SSOD sampler ($h/\delta = 1$) doubles the events generated with regard to the RQ sampler ($h/\delta = 0$). RQH sampling strategy reports intermediate values of generated events, generating a lesser amount of events with the reduction of h/δ.

To shed light on the advantages of RQH sampler with regard to SSOD or RQ consider the following example.

Example 2.1 Consider a SSOD, RQ and RQH sampler which are configured so that the steady-state error does not exceed $e_{ss} = 0.15$. For SSOD and RQH samplers a noise of 0.1 peak-to-peak amplitude is to be avoided. Thus, according to the provided guidelines, the parameter h of the RQH sampler is selected slightly higher than the noise amplitude, $h = 0.12$, and, δ is obtained from Eq. (2.2), $\delta = 2e_{ss} - h = 0.18$. For SSOD and RQ samplers, δ is determined by e_{ss}, being for SSOD $\delta = e_{ss} = 0.15$ and for RQ $\delta = 2e_{ss} = 0.3$.

To show the behavior of these samplers, a simulation has been conducted considering a sinusoidal input with unitary amplitude. In the initial 2 s the input to the sampler does not contain noise.

From the simulated results, it can be noted how the switching thresholds of the samplers and the quantification differ from one another. This is caused by the e_{ss} specification, producing different δ values for each sampler. When sampling a signal without noise, the RQ sampler produces the least number of events. Nonetheless, burst of events are generated when considering noisy inputs. On the other hand, SSOD sampler does not generate extra events due to noise, however, it generates the most events in the absence of noise due to the lowest switching thresholds. Finally, RQH sampler does not generate events due to noise and, it generates a lesser amount of events compared with the SSOD sampler. Therefore, this sampling strategy can effectively reduce communication output.

From this example it can be extracted that RQH offers a reduction in communication output compared to RQ and SSOD sampling techniques. Despite the shown benefits, the possible effects on the robustness and overall closed-loop performance must be evaluated. This issue will be studied in the following section.

2.4 Robustness Against Oscillation Induced by RQH

The robustness against limit cycle oscillations is strongly influenced by the sampling strategy. Among the different methods for characterizing this robustness, the Describing Function (DF) has been proven successfully for this kind of samplers [19, 24]. This method is applicable as long as high-order harmonics are sufficiently attenuated by the linear part of the system, which for the context of this chapter, this hypothesis holds. In further chapters, the influence of high order harmonics will be evaluated.

In order to the DF method to be applied, the block scheme in Fig. 2.1 can be replaced by a more compact one with two blocks. The first block gathers the linear elements in the open-loop transfer function $G_{ol}(s) = C(s)G(s)$, where $G(s) = G_p(s)e^{-t_d s}$ and the second block contains the RQH sampling law and holds the last value sampled. The combination of the event generation block and the ZOH results in a non-linearity, which can be, therefore, studied through the DF method.

DF theory states that sustained oscillations take place in loops as the described above if:

$$G_{ol}(j\omega) = -\frac{1}{\mathcal{N}}, \quad \forall\omega, \tag{2.6}$$

where $\mathcal{N}$ is the describing function of the non-linearity. Equation (2.6) can be visualized in the Nyquist plane, if the frequency response representation of $G_{ol}(j\omega)$ and $-\frac{1}{\mathcal{N}}$ do not intersect limit cycle oscillations will be avoided. For the RQH sampling strategy, the DF equation including the characteristic parameters h and δ, is given by the following expression (see Appendix B.1):

$$\begin{aligned}\mathcal{N}(A,h) = \frac{2\delta}{A\pi}\Bigg[&\sum_{k=1}^{m}\sqrt{1-\left(\frac{\delta}{A}\left(k+\frac{h}{2\delta}-\frac{1}{2}\right)\right)^2}+\\ &\sum_{k=m+1}^{2m}\sqrt{1-\left(\frac{\delta}{A}\left(2m-k-\frac{h}{2\delta}+\frac{1}{2}\right)\right)^2}\Bigg]-j\frac{2hm\delta}{A^2\pi},\end{aligned} \tag{2.7}$$

where A is the amplitude of the oscillation and m is the number of thresholds crossed defined by $m = \left\lfloor \frac{A}{\delta} - \frac{h}{2\delta} + \frac{1}{2} \right\rfloor$. The particular expressions for SSOD and RQ describing functions can be deduced from this equation by substituting $h = \delta$ and $h = 0$ respectively. However, the focus will be on the intermediate cases found within $h \in]0, \delta[$, due to the aforementioned reasons.

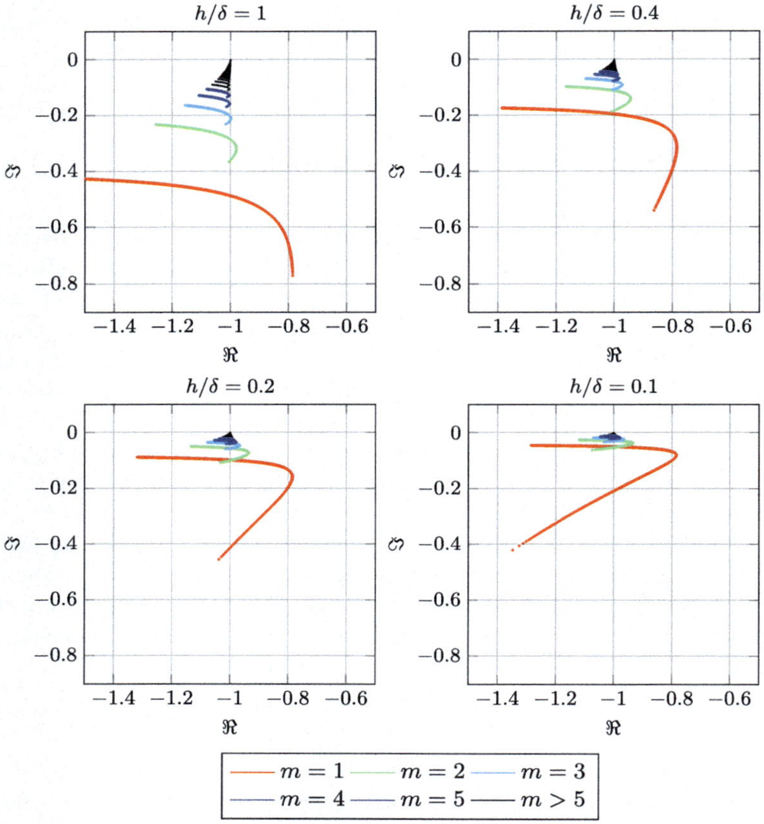

Fig. 2.2 Shapes of $-1/\mathcal{N}$ for RQH samplers with different values of h/δ. (Reprinted from IET Control Theory & Applications, Vol 14, No. 15, Miguel-Escrig, O., & Romero-Pérez, J. A., *Regular quantisation with hysteresis: a new sampling strategy for event-based PID control systems*, 2163–2175, Copyright (2020), with permission from John Wiley and Sons)

The shapes of $-1/\mathcal{N}$ for different values of h/δ is depicted in Fig. 2.2. Several branches compose the locus of $-1/\mathcal{N}$, each of them given for a value of m. These branches tend to fold, moving towards the abscissa axis as h/δ decreases. In Fig. 2.2 it is also shown the SSOD case by including $h/\delta = 1$, which was studied, together with the RQ sampler, by the authors in [24]. In that work, some measures were defined to characterize the robustness against limit cycle oscillations. However, those measures cannot be extended to the intermediate cases defined by the RQH, thus, new robustness measures are necessary.

These new measures can be defined using a similar concept as the classical gain and phase margins. It is defined as $\gamma_{h/\delta}$ the gain margin to the non-linearity, which is the necessary gain that makes G_{ol} intersect with $-1/\mathcal{N}$:

$$\gamma_{h/\delta} = \frac{\left|-\frac{1}{\mathcal{N}(A_\gamma)}\right|}{|G_{ol}(j\omega_\gamma)|}, \tag{2.8}$$

where ω_γ and A_γ are the frequency and amplitude for which the quotient $\frac{\left|-\frac{1}{\mathcal{N}(A)}\right|}{|G_{ol}(j\omega)|}$ is minimum while fulfilling the condition $\arg(G_{ol}(j\omega)) = \arg\left(-\frac{1}{\mathcal{N}(A)}\right)$. Rewriting this in a compact manner:

$$(\omega_\gamma, A_\gamma) = \arg\min_{(\omega,A)} \left(\frac{\left|-\frac{1}{\mathcal{N}(A)}\right|}{|G_{ol}(j\omega)|} : \arg(G_{ol}(j\omega)) = \arg\left(\frac{-1}{\mathcal{N}(A)}\right)\right). \tag{2.9}$$

Conversely, Is is defined as $\Phi_{h/\delta}$ the phase margin to the non-linearity, which is the minimum required phase by G_{ol} to intercept the DF traces. That is:

$$\Phi_{h/\delta} = \arg(G_{ol}(j\omega_\Phi)) - \arg\left(-\frac{1}{\mathcal{N}(A_\Phi)}\right), \tag{2.10}$$

where ω_Φ and A_Φ are the frequency and amplitude for which the difference $\arg(G_{ol}(j\omega)) - \arg\left(-\frac{1}{\mathcal{N}(A)}\right)$ is minimum and the condition $\left|-\frac{1}{\mathcal{N}(A)}\right| = |G_{ol}(j\omega)|$ is fulfilled. Rewriting this in a compact manner:

$$(\omega_\Phi, A_\Phi) = \arg\min_{(\omega,A)} \left(\arg(G_{ol}(j\omega)) - \arg\left(-\frac{1}{\mathcal{N}(A)}\right) : \left|-\frac{1}{\mathcal{N}(A)}\right| = |G_{ol}(j\omega)| \right). \tag{2.11}$$

Remark 2.2 As proved in [19, 24] for SSOD-PI and RQ-PI loops, using controllers tuned with phase margins greater than 45°, no intersection is observed between G_{ol} and the branch corresponding to $m = 1$, and therefore, this guarantees no intersections with other branches. In practice, it was proved that limit cycle oscillations can be avoided by assuring no intersection between G_{ol} and the branch of $-1/\mathcal{N}$ corresponding to $m = 1$. Hence, the proposed margins $\gamma_{h/\delta}$ and $\Phi_{h/\delta}$ will be measured with respect to the $m = 1$ branch of the RQH sampler.

The proposed $\gamma_{h/\delta}$ and $\Phi_{h/\delta}$ margins are presented on Fig. 2.3 in the Nyquist and Nichols diagrams. A given system and RQH sampler have been used for illustrative purposes. Note that only the branch corresponding to $m = 1$ of $-1/\mathcal{N}$ has been represented. Nichols chart facilitates the visualization of the margins since they are

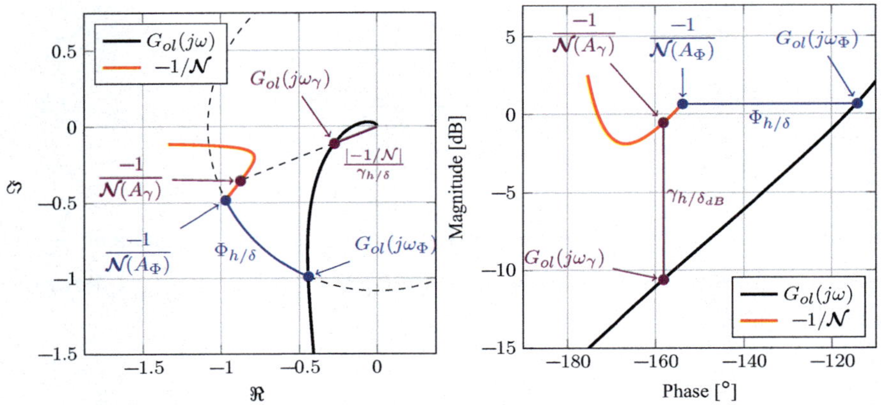

Fig. 2.3 Gain and phase margins to the non-linearity ($\gamma_{h/\delta}$, $\Phi_{h/\delta}$) for a given open-loop transfer function G_{ol} in Nyquist and Nichols diagrams. (Reprinted from IET Control Theory & Applications, Vol 14, No. 15, Miguel-Escrig, O., & Romero-Pérez, J. A., *Regular quantisation with hysteresis: a new sampling strategy for event-based PID control systems*, 2163–2175, Copyright (2020), with permission from John Wiley and Sons)

presented as vertical and horizontal lines, corresponding to the minimum distances from the open-loop transfer function to the RQH's inverse negative of $\mathcal{N}$.

Remark 2.3 As proposed in [30], the criterion for preventing limit cycle oscillations stems from an extension of the Nyquist stability criterion, treating the describing function as a generalized gain. This extension dictates that the entire Nyquist curve of the open-loop transfer function must encircle, in an anticlockwise manner, the critical points $(-1/\mathcal{N}, 0)$ a number of times equal to the count of poles with positive real parts in G_{ol}. Consequently, if neither the plant $G_p(s)$ nor the controller $C(s)$ possesses poles with positive real parts, the Nyquist curve should avoid encircling the negative inverse of the describing function to avert limit cycles. This principle allows defining $\gamma_{h/\delta}$ and $\Phi_{h/\delta}$ margins when encirclements are not required for stability. Thus, these margins are not applicable to plants containing positive real poles.

Remark 2.4 The proposed margins prevent the intersection between $-1/\mathcal{N}$ and $G_{ol}(s)$ which may not be easily prevented due to modeling errors, variations in the plant dynamics or the influence of the network delay. Similarly to classical margins, the proposed margins will also absorb system variations for a certain range. The plant model $G_p(s)$ strongly influences the magnitude of the admissible variations, thus, it must be studied specifically. The communication delay also influences the robustness, and its effects will be studied in subsequent section.

2.4.1 *Network Communication Delay Effect on the Robustness Margins*

The communication delay is taken into account in the term $e^{-t_d s}$ in the open-loop transfer function. This parameter t_d can suffer variations which may lead to a decrease in robustness against limit cycle oscillations. A study has been conducted to asses its influence on the proposed margins. Consider a variation on t_d: $t'_d = t_d \pm \alpha$, resulting in the following open-loop transfer function:

$$G'_{ol}(j\omega) = G_{ol}(j\omega)e^{\mp j\omega\alpha}. \tag{2.12}$$

Which can be expressed in magnitude and phase terms as:

$$\begin{cases} |G'_{ol}(j\omega)| = |G_{ol}(j\omega)| \\ \arg\{G'_{ol}(j\omega)\} = \arg\{G_{ol}(j\omega)\} \mp \omega\alpha \end{cases}. \tag{2.13}$$

Interpreting this equation in the Nichols diagram, it can be seen that variations on the communication delay result in horizontal movements of $G_{ol}(j\omega)$. This movement results in lower $\gamma_{h/\delta}$ and $\Phi_{h/\delta}$ as t'_d increases. The opposite effect is observed if t'_d decreases. For plants where this variation is critical, it is advised to consider the worst case scenario for the communication delay.

Remark 2.5 Depending on plant dynamics, gathered in $G_p(s)$, network delay may be negligible and, therefore, so will be its influence on the robustness margins. In those cases, it can be omitted from $G_{ol}(s)$.

2.5 Evaluation of Classical Tuning Methods Through the Proposed Margins

An analysis study has been conducted in which the proposed margins have been applied to study the degree of robustness of RQH-PI loops as in Fig. 2.1. The PI controllers are considered under the form:

$$C(s) = K_p\left(1 + \frac{1}{T_i s}\right), \tag{2.14}$$

and the tuning methods selected for the study have been Ziegler-Nichols [27], Cohen-Coon [28] and AMIGO [29]. Using these techniques, PI controllers have been tuned for the batch of models presented in (2.15), which gather a wide range of behaviors found in actual real systems. To tune the controllers, the responses of the models in the batch were approximated by First Order Plus Time Delay (FOPTD) models. The proposed robustness margins $\gamma_{h/\delta}$ and $\Phi_{h/\delta}$ have been measured in all cases.

$$
\begin{aligned}
&G(s) = \frac{e^{-s}}{(Ts+1)^2}, \\
&T = 0.01, 0.02, 0.05, 0.1, 0.2, 0.3, 0.5, 0.7, 1, \\
&\quad 1.3, 1.5, 2, 4, 6, 8, 10, 20, 50, 100, 200, 500 \\
\\
&G(s) = \frac{1}{(s+1)(Ts+1)^2}, \\
&T = 0.05, 0.1, 0.2, 0.5, 2, 5, 10 \\
\\
&G(s) = \frac{1}{(s+1)^n}, \\
&n = 3, 4, 5, 6, 7, 8 \\
\\
&G(s) = \frac{1}{(s+1)(\alpha s+1)(\alpha^2 s+1)(\alpha^3 s+1)}, \\
&\alpha = 0.1, 0.2, 0.3, 0.4, 0.5, 0.6, 0.7, 0.8, 0.9 \\
\\
&G(s) = \frac{Te^{-L_1 s}}{(T_1 s+1)(Ts+1)}, \\
&T_1 + L_1 = 1, \quad T = 1, 2, 5, 10 \\
&L_1 = 0.01, 0.02, 0.05, 0.1, 0.3, 0.5, 0.7, 0.9, 1 \\
\\
&G(s) = \frac{1-\alpha s}{(s+1)^3}, \\
&\alpha = 0.1, 0.2, 0.3, 0.4, 0.5, 0.6, 0.7, 0.8, 0.9, 1, 1.1 \\
\\
&G(s) = \frac{1}{(s+1)((sT)^2 + 1.4sT + 1)}, \\
&T = 0.1, 0.2, 0.3, 0.4, 0.5, 0.6, 0.7, 0.8, 0.9, 1
\end{aligned}
\tag{2.15}
$$

The resulting open-loop transfer functions, $G_{ol}(s) = C(s)G(s)$, from applying the designs based on the aforementioned tuning methods, are presented in Nyquist diagram in Fig. 2.4. In that figure, the traces of $-1/\mathcal{N}$ for $m = 1$ using different ratios h/δ are also presented.

From the presented figures it can be extracted that Ziegler-Nichols and Cohen-Coon methods cannot be applied to all type of processes and samplers since some of them present an intersection between $G_{ol}(s)$ and $-1/\mathcal{N}$, resulting in oscillatory behaviors. On the other hand, AMIGO method successfully avoids intersection with DF traces for all the studied cases, therefore, no limit cycle oscillations are expected in RQH-PI loops tuned with this method.

In general lines, due to this dependence on the tuning method used, the proposed robustness margins must be checked to assure a non-oscillatory response. Once obtained, the admissible variations on the plant model can be deduced and assess whether a re-tuning is necessary or not to avoid limit cycle oscillations.

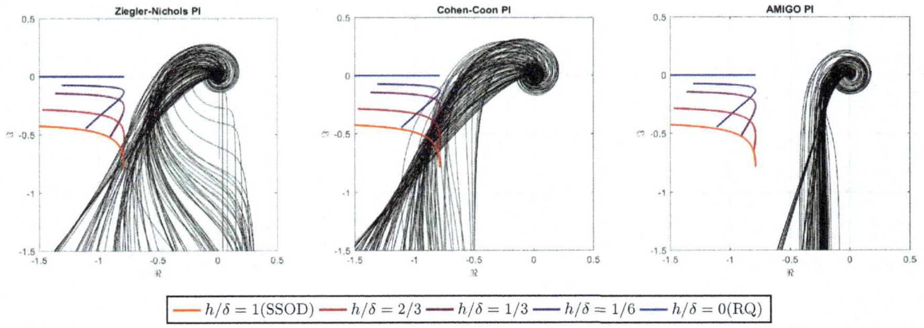

Fig. 2.4 Nyquist plots of the presented batch of processes with the specified controllers and the traces of $-1/\mathcal{N}$ for $m = 1$ with different values of h/δ. (Reprinted from IET Control Theory & Applications, Vol 14, No. 15, Miguel-Escrig, O., & Romero-Pérez, J. A., *Regular quantisation with hysteresis: a new sampling strategy for event-based PID control systems*, 2163–2175, Copyright (2020), with permission from John Wiley and Sons)

2.5.1 Influence of h/δ on the Robustness Margins

The proposed margins $\gamma_{h/\delta}$ and $\Phi_{h/\delta}$ have been measured on the precedent systems for different values of the ratio h/δ.

The measured results indicate that AMIGO tuning rule provides positive values for the margins across the studied processes and for all ratios h/δ. In contrast, controllers tuned according to Ziegler-Nichols and Cohen-Coon methods would provide oscillatory responses, concretely when using samplers with a high value of h/δ. This is remarkable specially considering that these methods provide reasonable values for the classical gain and phase margins. Among the studied methods, Cohen-Coon is to be avoided since limit cycle oscillations are expected for almost every process when using $h/\delta \geq 0.4$.

Generalizing from the precedent study, it can be deduced that a reduction in the ratio h/δ results in increased $\Phi_{h/\delta}$ and $\gamma_{h/\delta}$ margins, which tend to the classical gain and phase margin as the hysteresis tends to 0.

2.5.2 Influence of Controller's Parameters on the Robustness Margins

As in the classical case, the controller's parameters have a notable influence on the proposed robustness margins. Its parameters change the position of $G_{ol}(s)$ with respect to $-1/\mathcal{N}$, altering the margins. The influence of such movements can be observed in the Nichols diagram. By modifying the proportional gain K_p or the integral time T_i, vertical or horizontal movements of $G_{ol}(s)$ are produced respectively.

Expressing the PI controller in Eq. (2.14) in terms of magnitude and phase:

$$|C(j\omega)| = K_p\sqrt{1+\left(\frac{1}{\omega T_i}\right)^2}; \quad \arg\{C(j\omega)\} = \arctan\left(-\frac{1}{\omega T_i}\right),$$

the influence of each of the parameters can be easily spotted.

Reducing K_p results in a vertical downwards displacement of $G_{ol}(j\omega)$, which implies an improvement of both margins. On the other hand, varying T_i results in both a vertical and horizontal movement of $G_{ol}(j\omega)$. By increasing T_i the open-loop transfer function moves to the right and downwards, increasing both margins.

These results set the guidelines for detuning PI controllers in case that limit cycle oscillations are observed. By reducing K_p and/or increasing T_i, not only the ad-hoc margins to the non-linearity are increased, but also the classical gain and phase margins, which behave in the same manner. Obviously, this detuning may result in slower closed-loop responses.

2.6 Simulation Examples

In this section, several examples are provided to illustrate the contents treated in this chapter.

Example 2.2 To show the influence of the ratio h/δ on the event generation in closed loop, illustrating the contents developed in Sect. 2.3.1, consider a process modeled by the following transfer function:

$$G_p(s) = \frac{e^{-0.3\,s}}{(s+1)(0.7\,s+1)}. \tag{2.16}$$

The network communication delay is considered to have $t_d = 0.15$ seconds o latency, resulting in the following plant model:

$$G(s) = \frac{e^{-0.45\,s}}{(s+1)(0.7\,s+1)}. \tag{2.17}$$

A PI controller according to Ziegler-Nichols tuning rule has been tunned with $K_p = 1.52$ and $T_i = 2.58$. This controller avoids limit cycle oscillations induced by any RQH sampler since it avoids intersection with the inverse negative of the traces of the DF for any ratio h/δ.

Consider for this system to have a noise peak-to-peak amplitude of 0.07 units and a constraint on the maximum steady-state error of $e_{ss} = 0.18$. Following the guidelines provided in Sect. 2.3, firstly, a SSOD sampler is selected with $\delta = 0.18$, which fulfills the provided specifications. In addition, with the same guidelines a RQH sampler with $h = 0.08$ and $\delta = 0.28$ is selected, fulfilling the noise and steady-state error requirements.

Two experiments, consisting of two unitary step changes at the reference input and an additional one at the disturbance input, have been performed with both samplers.

In both cases similar responses are obtained, reporting settling times around 10 seconds and no remarkable differences regarding the overshoot. However, a major difference is observed in the number of generated events (n_{ev}). While the SSOD generates 31 events the RQH sampler only needs 22, showing an effective reduction in the number of generated events without altering significantly the system performance.

It is worth remarking that the quantization output δ directly affects the control action bumps, defined by $\delta_u = K_p\delta$, which, for samplers chosen according to the provided guidelines, would be greater for RQH than SSOD samplers.

Example 2.3 In this example, the usefulness of the describing function approach for predicting limit cycle oscillations in RQH-PI loops and the influence of the ratio h/δ in their appearance will be shown.

The process for this example is modeled by:

$$G_p(s) = \frac{1 - 0.1\,s}{(s+1)^3}, \tag{2.18}$$

and the latency is considered negligible with regard to system's dynamics. Using Cohen-Coon tuning rules a PI controller with $K_p = 1.612$ and $T_i = 1.938$ has been tuned.

Consider also a peak-to-peak noise amplitude of 0.03 units and a maximum steady-state error of $e_{ss} = 0.1$. According to these specifications, two RQH samplers are designed. Focusing on the first sampler, an hysteresis $h = 0.06$ is chosen and the quantification $\delta = 0.14$ matching the e_{ss} requirement, resulting in a sampler with ratio $h/\delta = 0.4286$. This sampler proposes a conservative approach with regard to event generation due to noise.

Testing in simulation against unitary changes in reference and disturbance inputs a limit cycle oscillation is observed in its temporal response.

Analyzing the system from a classical point of view, the crossover margins are $\gamma_{cp} = 6.5dB$ and $\Phi_{cg} = 30.37°$, which are large enough for most of continuous control applications. However, using the developed margins, it is obtained $\gamma_{h/\delta} = -0.4dB$ and $\Phi_{h/\delta} = -1.98°$ margins, meaning that an intersection between $G_{ol}(j\omega)$ and $-1/\mathcal{N}$ already exists.

Focusing on the second sampler, an hysteresis $h = 0.04$ and $\delta = 0.16$ to match the e_{ss} requirement have been chosen, resulting in a ratio $h/\delta = 0.25$, lower than in the precedent case. Analyzing the system with this sampler, a gain $\gamma_{h/\delta} = 1.15dB$ and phase $\Phi_{h/\delta} = 5.58°$ margins have been obtained, indicating that the open-loop transfer function does not intersect with the traces of the inverse negative of DF, and therefore, that no limit cycle oscillations will take place.

Simulations on this system show the same result, under the same conditions than the previous sampler. As expected, no limit cycle oscillations are observed, showing that the reduction on the ratio h/δ of the sampler allows lowering the robustness requirements to be fulfilled to avoid limit cycle oscillations.

Example 2.4 Modeling errors can appear in the industrial context due to multiple causes. Some of the most frequent are noisy measurement, scarce data for identification or hidden non-linear behaviors. The robustness margins, generally speaking, serve to absorb these errors and possible plant's dynamics variations. This example illustrates the role of the proposed margins in this sense.

A process nominal model is given by the transfer function:

$$G(s) = \frac{e^{-0.2\,s}}{(s+1)^4}, \tag{2.19}$$

whose gain, delay and poles have been obtained with a reliability of $\pm 5\%$. AMIGO tuning rule has been used to tune a PI controller with $K_p = 0.281$ and $T_i = 2.41$. A RQH sampler with ratio $h/\delta = 1/3$ has been considered.

Worst and best case scenarios due to uncertainties ($G_w(s)$ and $G_b(s)$ respectively), from the robustness perspective, have been obtained to be:

$$G_w(s) = \frac{1.05e^{-0.21\,s}}{(s+0.95)^4} \quad \text{and} \quad G_b(s) = \frac{0.95e^{-0.19\,s}}{(s+1.05)^4}. \tag{2.20}$$

The proposed margins have been measured for each case, resulting in $\gamma_{h/\delta}$ =9.7 dB and $\Phi_{h/\delta}$ =42.8° for the worst case, $\gamma_{h/\delta}$ =12.3 dB and $\Phi_{h/\delta}$ =48.3° for the nominal model, and, $\gamma_{h/\delta}$ =14.8 dB and $\Phi_{h/\delta}$ =52° for the best case.

Considering these data, the obtained AMIGO controller can be considered to prove a robust behavior against limit cycle oscillations induced by the RQH sampler even in the presence of uncertainties in the model.

If the structure of the model is known, the maximum admissible variation of each uncertain parameter can also be obtained. For this example, process gain can increase until 4.11, the delay until 6.98 seconds and the poles can be reduced to 0.75.

Quick summary:

- ☐ RQH sampler has been introduced as an alternative to RQ or SSOD sampling strategies for event-based PI control systems consisting on an intermediate solution.
- ☐ The proposed RQH reduces the number of generated events to perform the control while avoiding event generation due to noise.
- ☐ Describing function technique has been used to assess the robustness against limit cycles induced by the proposed sampling strategy.
- ☐ An ad-hoc gain and phase margins against limit cycle induction has been defined.

- ☐ Assessment of classical tuning methods for continuous systems (Ziegler-Nichols, Cohen-Coon, AMIGO) under the proposed sampling strategy with the defined margins has been performed.
- ☐ A study revealing the potential induction of limit cycle oscillations despite reasonable classical gain and phase margins have been carried out.
- ☐ The impact of controller parameters, network delay, and model uncertainty on the newly defined margins has been evaluated, observing similar behaviors than for classical margins.
- ☐ Guidelines have been provided for selecting sampler parameters based on noise and steady-state error specifications.

References

1. Lunze J (2015) Event-based control: introduction and survey. In: Miskowicz M (ed) Event-Based control and signal processing. CRC Press, Boca Raton, pp 3–20
2. Dotoli M, Fay A, Miśkowicz M, Seatzu C (2019) An overview of current technologies and emerging trends in factory automation. Int J Prod Res 57(15–16):5047–5067
3. Aranda-Escolástico E, Guinaldo M, Heradio R, Chacón J, Vargas H, Sánchez J, Dormido S (2020) Event-based control: a bibliometric analysis of twenty years of research. IEEE Access 8(March):47188–47208
4. O'Dwyer A (2006) Handbook of PI and PID controller tuning rules, 2nd ed. Imperial College Press
5. Samad T (2017) A survey on industry impact and challenges thereof. IEEE Control Syst 37(1):17–18
6. Maxim A, Copot D, Copot C, Ionescu CM (2019) The 5w's for control as part of industry 4.0: why, what, where, who, and when-A PID and MPC control perspective. Inventions 4(1)
7. Årzén K-E (1999) A simple event-based PID controller. In: Proceedings of the 14th IFAC world congress, vol 18, pp 423–428
8. Durand S, Marchand N (2009) An event-based PID controller with low computational cost. In: 8th international conference on sampling theory and applications (SampTA'09), pages Special–session
9. Durand S, Marchand N (2009) Further results on event-based PID controller. In: Control conference (ECC), 2009 European. IEEE, pp 1979–1984
10. Vasyutynskyy V, Kabitzsch K (2010) A comparative study of PID control algorithms adapted to send-on-delta sampling. In: 2010 IEEE international symposium on industrial electronics (ISIE), pp 3373–3379. IEEE
11. Vasyutynskyy V, Kabitzsch K (2009) Time constraints in PID controls with send-on-delta. IFAC Proc Vol 42(3):48–55
12. Hsia TC (1974) Analytic design of adaptive sampling control law in sampled-data systems. IEEE Trans Autom Control 19(1):39–42
13. Dormido S, Sánchez J, Kofman E (2008) Muestreo, control y comunicacián basados en eventos. Revista Iberoamericana de Automåatica e Informåatica Industrial RIAI 5(1):5–26
14. Ploennigs J, Vasyutynskyy V, Kabitzsch K (2010) Comparative study of energy-efficient sampling approaches for wireless control networks. IEEE Trans Ind Inf 6(3):416–424
15. Beschi M, Dormido S, Sánchez J, Visioli A (2012) Characterization of symmetric send-on-delta PI controllers. J Process Control 22(10):1930–1945

16. Beschi M, Dormido S, Sánchez J, Visioli A (2014) Tuning of symmetric send-on-delta proportional-integral controllers. IET Control Theory Appl 8:248–259(11)
17. Pawlowski A, Beschi M, Guzmán JL, Visioli A, Berenguel M, Dormido S (2016) Application of SSOD-PI and PI-SSOD event-based controllers to greenhouse climatic control. ISA Trans 65:525–536
18. Romero-Pérez J-A, Sanchis-Llopis R, Peñarrocha-Alós I (2014) A simple rule for tuning event-based PID controllers with symmetric send-on-delta sampling strategy. In: Proceedings of the 2014 IEEE emerging technology and factory automation (ETFA), pp 1–8, Sept 2014
19. Romero-Pérez J-A, Sanchis-Llopis R (2016) A new method for tuning PI controllers with symmetric send-on-delta sampling strategy. ISA Trans 64:161–173
20. Ruiz Á, Beschi M, Visioli A, Dormido S, Jiménez JE (2017) A unified event-based control approach for FOPTD and IPTD processes based on the filtered Smith predictor. J Franklin Inst 354(2):1239–1264
21. Sánchez J, Guinaldo M, Visioli A, Dormido S (2018) Identification of process transfer function parameters in event-based PI control loops. ISA Trans 75:157–171
22. Miguel-Escrig O, Romero-Pérez J-A, Sanchis-Llopis R (2020) Tuning PID controllers with symmetric send-on-delta sampling strategy. J Franklin Inst 357(2):832–862
23. Miguel-Escrig O, Romero-Pérez J-A, Sanchis-Llopis R (2018) New robustness measure for a kind of event-based PID. IFAC-Papers online 51(4):781–786
24. Romero-Pérez J-A, Sanchis-Llopis R (2017) Tuning and robustness analysis of event-based PID controllers under different event generation strategies. Int J Control, pp 1–38
25. Åström KJ, Hägglund T (2006) Advanced PID Control. Systems, and Automation Society, ISA-The Instrumentation
26. Sun L, Li D, Lee KY (2016) Optimal disturbance rejection for PI controller with constraints on relative delay margin. ISA Trans 63:103–111
27. Ziegler JG, Nichols NB (1942) Optimum settings for automatic controllers. Trans ASME 64(11)
28. Cohen GH, Coon GA (1953) Theoretical consideration of retarded control. Trans ASME 75:827–834
29. Åström KJ, Hägglund T (2004) Revisiting the Ziegler-Nichols step response method for PID control. J Process Control 14(6):635–650
30. Jean-Jacques ES, Weiping Li (1991) Applied nonlinear control. Prentice Hall

Chapter 3
Characterization of Limit Cycle Oscillations Induced by Fixed Threshold Samplers

3.1 Introduction

Recently, Event-Based Control (EBC) is gaining increasing popularity as a control alternative. This paradigm shift is driven by the emergence of Industry 4.0, a novel industrial framework. The demands of this new communication standard emphasize the need for efficient data flow across industrial networks, thereby encouraging the adoption of EBC algorithms.

Numerous researchers have applied EBC to established control algorithms, particularly to the PID algorithm. The PID algorithm, known for its simplicity and reliability, is widely used in the industry [1]. Furthermore, its relevance in the context of Industry 4.0 has been emphasized in [2].

Årzén's seminal work [3] stands as one of the initial references to the Event-Based PID control algorithm. This contribution demonstrated the algorithm's utility in minimizing CPU usage for process control, without substantial impact on performance. The study identified crucial aspects of EBC, particularly addressing errors in the calculation of the integral and derivative terms when the time between samples increases.

Subsequent efforts have been dedicated to address these issues, specially focusing on the calculation of the integral action. Noteworthy contributions in this regard come from Durand [4, 5] and Vasyutynskyy [6, 7]. Additionally, works like [8] demonstrate the effectiveness of these algorithms when implemented using the standard IEC 61499 [9], which facilitates the deployment of distributed control systems.

The sampling strategy plays a paramount role together with the control algorithm. The sampling strategy of choice must be responsible for triggering events upon detecting significant changes in the state of the controlled system. Fixed

---

Adapted with changes in the text from IEEE Access, Vol 10, Miguel-Escrig, O., Romero-Pérez, J. A., Sánchez-Moreno, J., Dormido, S., *Characterization of limit cycle oscillations induced by fixed threshold samplers*, 62581-62596, Copyright (2022), under license CC BY 4.0 https://creativecommons.org/licenses/by/4.0/.

O. Miguel-Escrig, *Event-Based PID Controllers with Fixed Threshold Sampling Strategies*, Springer Theses, https://doi.org/10.1007/978-3-031-62505-3_3

Threshold Samplers (FTS) can be found as the most common for its ease of implementation. Examples of such sampling strategies include Regular Quantization (RQ), Symmetric-Send-On-Delta (SSOD) [10], or Regular Quantization with Hysteresis (RQH) [11], representing an intermediate approach.

Numerous studies have delved into incorporating these strategies into the control loop. While some have utilized them for identification purposes [12–14], a more significant focus has been on employing them for process control [15–17]. To ensure effective performance, several tuning rules and procedures have been proposed [18–21].

The majority of the mentioned studies address a drawback associated with these types of FTS, namely their potential to induce limit cycle oscillations. In the context of EBC, these oscillations detrimentally impact loop performance, wearing out actuators, and overloading communication networks. However, when considering these oscillations from the perspective of relay control principles, an oscillatory state can be considered an expected outcome.

This chapter focuses on characterizing the limit cycle oscillations induced by various types of FTS. For this purpose, an analysis tool called Dual Input Describing Function (DIDF) [22], a variant of the Describing Function (DF), is employed. The study involves parameterizing the FTS and obtaining the general DIDF to assess whether sustained oscillations may be induced in a given loop.

3.2 Problem Statement

Selecting an appropriate sampler or adapting it to specific requirements can be a challenging task, particularly when considering that just by quantifying limit cycle oscillations may be induced.

Moreover, as the complexity of the sampler increases, the oscillations may become challenging to anticipate. For instance, consider the oscillation depicted in Fig. 3.1 generated by a particular FTS.

In this figure, a substantial disparity between the input signal $x(\phi)$ and the output $\bar{x}(\phi)$ of the FTS is evident. It can be observed a difference in the quantization step, which surpasses the increment in the signal, an asymmetry in the input and output of the FTS, and a different number of crossed levels between upwards (four) and downwards (three) directions.

According to the relationship between input (x) and output ($\bar{x}$) of a generalized FTS, presented in Fig. 3.2, the aforementioned traits can be parametrized.

In this figure, the parameters defining a FTS are presented. These parameters are: the detection width δ, the output quantification ϵ, the hysteresis h and the asymmetries on the detection $\Delta\delta$ and output $\Delta\epsilon$. Note that the asymmetries are measured from the center of the rectangle $\delta \times \epsilon$, being the hysteresis h centered vertically on that rectangle. It is considered $\Delta\epsilon \in [-\epsilon/2, \epsilon/2]$ and $\Delta\delta \in [-\delta/2, \delta/2]$ without loss of generalization

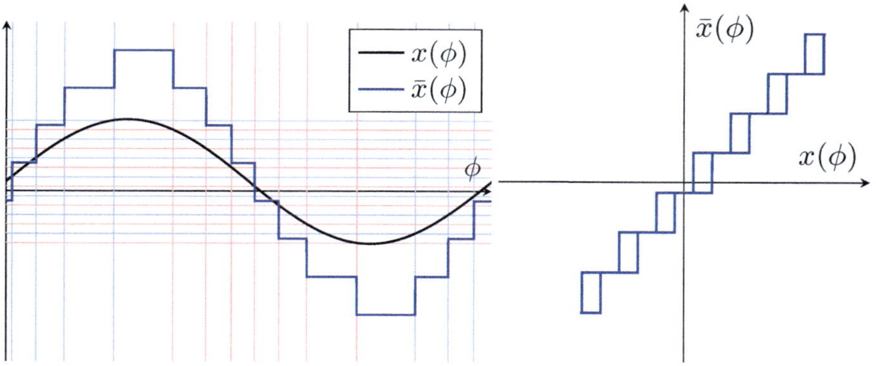

Fig. 3.1 On the left image, sinusoidal oscillation sampled with a given FTS. On the right: input-output relationship of the chosen FTS (parameters: $\Delta\delta/\delta = 1/2$, $\Delta\epsilon/\epsilon = 1/4$, $\epsilon/\delta = 2$, $h/\delta = 1/2$, explained below). (Reproduced from IEEE Access, Vol 10, Miguel-Escrig, O., Romero-Pérez, J. A., Sánchez-Moreno, J., Dormido, S., *Characterization of limit cycle oscillations induced by fixed threshold samplers*, ©IEEE 2022)

Fig. 3.2 Relationship between input and output of a generic FTS. (Reproduced from IEEE Access, Vol 10, Miguel-Escrig, O., Romero-Pérez, J. A., Sánchez-Moreno, J., Dormido, S., *Characterization of limit cycle oscillations induced by fixed threshold samplers*, ©IEEE 2022)

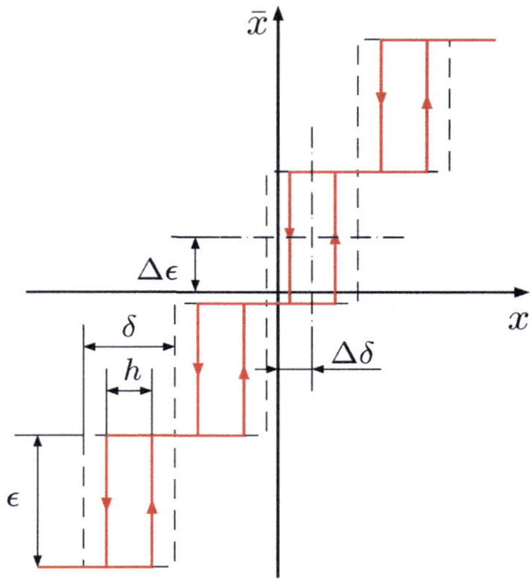

Through these parameters some conclusions can be extracted about sampler and system behavior. For instance, if the sampler has $\Delta\epsilon = \pm\epsilon/2$ will contain a dead zone, or the output would be symmetric if $\Delta\delta = \pm\delta/2$, or can implicitly include a gain depending on the relationship between ϵ and δ.

Due to the generalization that this FTS provides, depending on the choice of parameters, the most common FTS sampelrs can be found. A summary of some of them can be found in Table 3.1.

Table 3.1 Summary of samplers and their parameter configuration

Sampler	Parameter configuration	Representation
RQ	$\Delta\delta = \delta/2, \Delta\epsilon = \epsilon/2$ $\epsilon = \delta, h = 0$	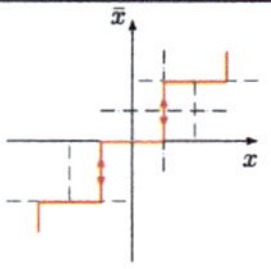
RQH [41]	$\Delta\delta = \delta/2, \Delta\epsilon = \epsilon/2$ $\epsilon = \delta$	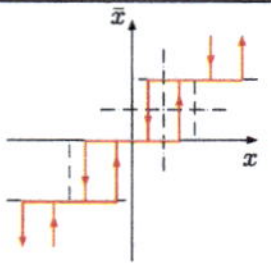
SSOD [10]	$\Delta\delta = \delta/2, \Delta\epsilon = \epsilon/2$ $\epsilon = \delta = h$	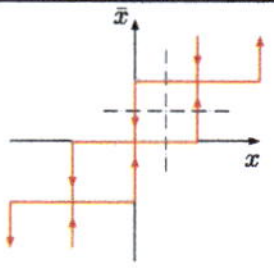
Input and/or Output biased quantifier	$\Delta\delta \neq \delta/2$ and/or $\Delta\epsilon \neq \epsilon/2$ $\epsilon = \delta$	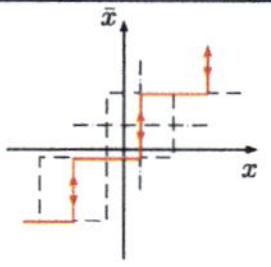
Multilevel ideal relay	$\Delta\delta = 0, \Delta\epsilon = 0$ $\epsilon = \delta = D/2, h = 0$	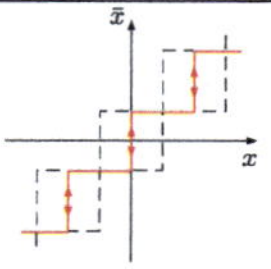
Multilevel relay with hysteresis	$\Delta\delta = 0, \Delta\epsilon = 0$ $\epsilon = \delta = h = D/2$	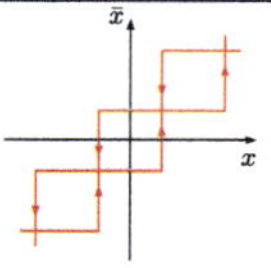

3.3 DIDF Approach

As aforementioned, a crucial phenomenon associated with using a FTS is its potential to induce limit cycle oscillations. To analyze such nonlinear behavior, various tools are available, with the Describing Function approach standing out for its ease of use and accuracy in the results obtained [23, 24].

The analyzed generic FTS encompasses a broad range of commonly used samplers, including SSOD [10], RQH [11], or RQ sampling, as demonstrated in the preceding section. Thus, by conducting an analysis based on the most generic case, a

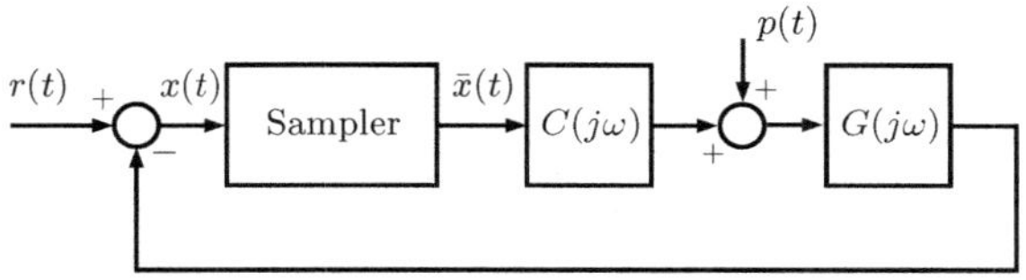

Fig. 3.3 Generic loop structure that allows the DF technique application. (Reproduced from IEEE Access, Vol 10, Miguel-Escrig, O., Romero-Pérez, J. A., Sánchez-Moreno, J., Dormido, S., *Characterization of limit cycle oscillations induced by fixed threshold samplers*, ©IEEE 2022)

diverse array of samplers can be characterized through a single methodology. Based on the parameters described before, the relationship between the input signal to the sampler $x(t)$ and its output $\bar{x}(t)$ is defined as follows:

$$\bar{x}(t) = \begin{cases} \Delta\epsilon - \frac{\epsilon}{2} + \epsilon(i+1) \text{ if } x(t) \geq \Delta\delta + \frac{h}{2} + \delta i \text{ and } \bar{x}(t^-) = \Delta\epsilon - \frac{\epsilon}{2} + \epsilon i, \quad i \in \mathbb{Z} \\ \Delta\epsilon - \frac{\epsilon}{2} + \epsilon(i-1) \text{ if } x(t) \leq \Delta\delta - \frac{h}{2} + \delta(i-1) \text{ and } \bar{x}(t^-) = \Delta\epsilon - \frac{\epsilon}{2} + \epsilon i \\ \Delta\epsilon - \frac{\epsilon}{2} + \epsilon i \qquad \text{if } x(t) \in \left[\Delta\delta - \frac{h}{2} + \delta(i-1), \Delta\delta + \frac{h}{2} + \delta i\right] \text{ and } \bar{x}(t^-) = \Delta\epsilon - \frac{\epsilon}{2} + \epsilon i \end{cases}$$

In order to apply the analysis methodology consider the generic loop presented in Fig. 3.3. The controller $C(j\omega)$ and the process $G(j\omega)$ blocks can be grouped in a single block $G_{ol}(j\omega) = C(j\omega)G(j\omega)$ representing the open-loop transfer function, gathering all the linear behavior.

Due to the inherent generalization introduced by the FTS, there may be potential biases in the sampler's output that cannot be mitigated through the loop and, as a result, this bias should be taken into account at the input of the non-linearity. Hence, the input to the sampler takes the form $x(t) = A\sin(\omega t) + B$. To encompass cases where a bias is present in the study, a variation of the DF known as Dual Input Describing Function (DIDF) is utilized. The DIDF assesses the sustainability of the oscillation and considers the bias in the analysis.

According to the DIDF theory, the condition for attaining a sustained oscillatory behavior, also called limit cycle oscillation, for the systems under study is given by:

$$G_{ol}(j\omega) = -\frac{1}{\mathcal{N}_A}, \quad \forall\omega, \tag{3.1}$$

being $\mathcal{N}_A$ the describing function of the sampler under study described by a non-linear element, in this case the FTS. Graphically, this condition translate to: limit cycle oscillations can appear if an intersection between the open-loop transfer function and the inverse negative of the describing function exists in the Nyquist diagram, then, limit cycle oscillations can be induced.

Besides to the condition for the sustainability of oscillations, presented in Eq. (3.1), an additional condition is required to evaluate the sustainability of the possible induced bias in the oscillation. Some of the conditions for the bias maintenance were described in [25] for some simple cases. These conditions also depend on the linear part of the system under study.

The validity of the presented conditions on that work were limit to the cases where disturbance and reference changes were not considered, i.e. $r(t) = 0$ and $p(t) = 0$, thus, these conditions need adjustment for the present context.

Consider the system scheme depicted in Fig. 3.3, and assume reference and disturbance inputs to be signals changing in a step-like manner. Also, for further generalization, consider the oscillation to contain a bias B. Studying the non-oscillatory terms of the elements in the loop:

$$\begin{aligned} x(\infty) &= r(\infty) - y(\infty) \\ B &= r(\infty) - G(0)[p(\infty) + B\mathcal{N}_B C(0)], \end{aligned}$$

where $r(\infty)$ and $p(\infty)$ are the final value of the step changes, $G(0)$ and $C(0)$ are the dcgain of the linear elements in the loop, B the bias and $\mathcal{N}_B$ the describing function for the bias sustainability. Elements in the previous expression can be arranged, obtaining the general condition for bias maintainability:

$$r(\infty) - p(\infty)G(0) = B(1 + \mathcal{N}_B G_{ol}(0)), \tag{3.2}$$

where $G_{ol}(0)$ is the product $G(0)C(0)$, corresponding to the open-loop transfer function.

The form of $G(s)$ and $C(s)$ is of paramount importance for the conditions of the maintenance of the bias, which may change according to their form. A complete study is presented in Table B.1 of Appendix B.3.

The expressions for the DIDF of the general FTS involving the bias maintainability and the oscillation maintenance are given respectively by (see Appendix B.2):

$$\begin{aligned} \mathcal{N}_B = &\frac{2\Delta\epsilon + \epsilon(m_{sup} - m_{inf} + 2i_0 - 1)}{2B} \\ &- \frac{\epsilon}{2\pi B} \sum_{k=-m_{inf}}^{m_{sup}-1} \left[\arcsin\left(\frac{1}{A}\left(\Delta\delta + \frac{h}{2} - B + \delta(k + i_0)\right)\right)\right. \\ &\left. + \arcsin\left(\frac{1}{A}\left(\Delta\delta - \frac{h}{2} - B + \delta(k + i_0)\right)\right)\right] \end{aligned} \tag{3.3}$$

for the bias, and:

$$\mathcal{N}_A = \frac{\epsilon}{\pi A} \sum_{k=-m_{inf}}^{m_{sup}-1} \left[\sqrt{1 - \left(\frac{1}{A}\left(\Delta\delta + \frac{h}{2} - B + \delta(k+i_0)\right)\right)^2} + \sqrt{1 - \left(\frac{1}{A}\left(\Delta\delta - \frac{h}{2} - B + \delta(k+i_0)\right)\right)^2} \right] - j\frac{\epsilon h}{\pi A^2}[m_{sup} + m_{inf}], \quad (3.4)$$

for the oscillation. In these expression, the term i_0 represents the initial or base level, around which the oscillations are studied. Besides, the terms m_{sup} and m_{inf} are the number of levels crossed in the upwards and downwards direction respectively counting from the base level i_0. These parameters can be analytically obtained according to the following expressions:

$$m_{sup} = \left\lfloor \frac{A + B - \Delta\delta - h/2 - \delta i_0}{\delta} \right\rfloor + 1, \quad m_{inf} = -\left\lfloor \frac{-A + B - \Delta\delta + h/2 - \delta i_0}{\delta} \right\rfloor - 1 \quad (3.5)$$

and:

$$i_0 = \begin{cases} \left\lfloor \dfrac{B - \Delta\delta - h/2}{\delta} \right\rfloor + 1 & \text{if } B \geq 0 \\ \left\lfloor \dfrac{B - \Delta\delta + h/2}{\delta} \right\rfloor & \text{if } B < 0 \end{cases}.$$

The DF can be expressed using a series of dimensionless ratios easing the analysis. These ratios are ϵ/A, ϵ/B, δ/A, δ/B, $\Delta\epsilon/\epsilon$, $\Delta\delta/\delta$ and h/δ; and some of them are bounded in value: $\Delta\epsilon/\epsilon \in [-1/2, 1/2]$, $\Delta\delta/\delta \in [-1/2, 1/2]$ and $h/\delta \in [0, 1]$.

The previous expression can be used to obtain the describing function of the cases presented in Table 3.1, precisely, the expressions of the dynamic DF ($\mathcal{N}_A$) considering $B = 0$ can be found for some of the most documented like RQ [24], RQH [11] or SSOD [23] for instance.

3.3.1 Characteristics Related to the Dynamic Part of the DIDF

As a consequence of the generality of the provided expression for $\mathcal{N}_A$, directly committing an analysis will result in a complex task. Thus, to limit the analysis effort, the dynamic part of the DIDF is studied in this section.

Property 3.1 (Different oscillatory modes for a given FTS) *One of the characteristic behaviors that FTS can present has already been depicted in Fig. 3.1, which is the*

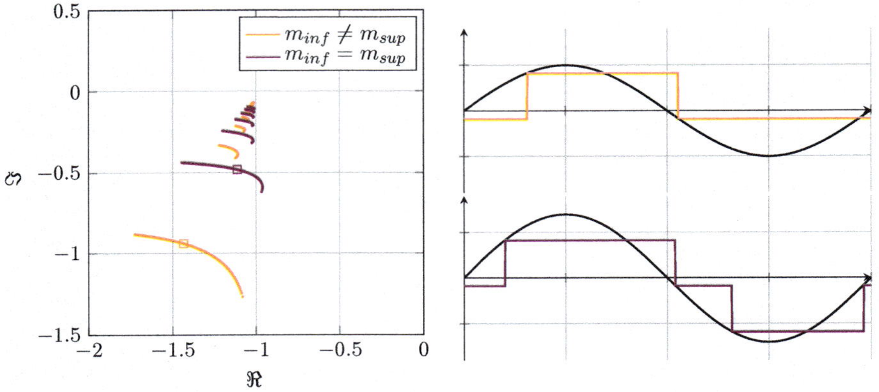

Fig. 3.4 $\mathcal{N}_A$ of a sampler with two series of traces regarding the levels crossed in each direction (Parameters used: $\Delta\epsilon/\epsilon = \Delta\delta/\delta = 1/3$, $\epsilon = \delta = h$). (Reproduced from IEEE Access, Vol 10, Miguel-Escrig, O., Romero-Pérez, J. A., Sánchez-Moreno, J., Dormido, S., *Characterization of limit cycle oscillations induced by fixed threshold samplers*, ©IEEE 2022)

possible induction of limit cycle oscillations that cross a different number of levels in the upwards an downwards direction counting from i_0, i.e. $m_{sup} \neq m_{inf}$. This phenomenon can be observed regardless of the analyzed FTS, which could also provide symmetric oscillations, i.e. $m_{sup} = m_{inf}$, for another value of the oscillation amplitude.

Studying this phenomenon it has been found that two "series" of alternate bands appear on the inverse negative of the dynamic DF, one representing the case $m_{sup} \neq m_{inf}$ and another the $m_{sup} = m_{inf}$ case. This potential difference between m_{sup} and m_{inf} can also be observed evaluating expressions in (3.5) for certain values.

In Fig. 3.4 this behavior has been illustrated. On the left image, the dynamic DF $\mathcal{N}_A$ (Eq. (3.4)) of a sampler with $\Delta\epsilon/\epsilon = \Delta\delta/\delta = 1/3$, $\epsilon = \delta = h$ and $B = 0$ has been depicted. The two series of traces have been represented in different colors. On those traces, a square has been marked, which correspond to the oscillatory behavior presented on the images on the right. Each of the observed oscillations present a different numbers of levels crossed, corresponding to its respective square on the Nyquist diagram.

Property 3.2 (Implicit sampler gain) *The ratio ϵ/δ does not explicitly appear in expressions (3.4) and (3.3). However, it is of great importance due to its influence on the placement of the inverse negative of the DF traces on the Nyquist diagram. This ratio holds the relationship between the detection input and the quantization output of the sampler, hence, it behaves to a certain extent as a gain.*

The effect of varying this gain is reflected in the DF traces as a radial shrink or expansion, consequence of increasing or decreasing this ratio, respectively. Hence, this ratio is of paramount importance regarding to the loop robustness, restricting

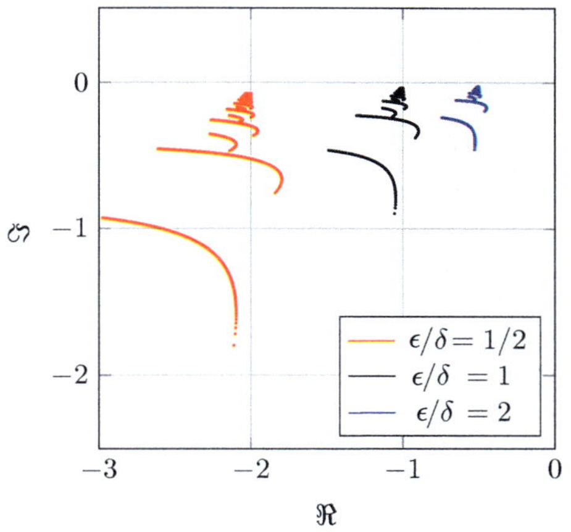

Fig. 3.5 Radial variance of $-1/\mathcal{N}_A$ with the ratio ϵ/δ (Parameters used: $\Delta\epsilon/\epsilon = 1/4$, $\Delta\delta/\delta = 1/3$ and $h/\delta = 1/2$). (Reproduced from IEEE Access, Vol 10, Miguel-Escrig, O., Romero-Pérez, J. A., Sánchez-Moreno, J., Dormido, S., *Characterization of limit cycle oscillations induced by fixed threshold samplers*, ©IEEE 2022)

the gain of the frequency response of the linear elements in the loop which behaves oppositely.

For a given sampler, defined by $\Delta\epsilon/\epsilon = 1/4$, $\Delta\delta/\delta = 1/3$ *and* $h/\delta = 1/2$, *the radial variation induced by the change in the ratio* ϵ/δ *has been presented in Fig. 3.5.*

Property 3.3 (Hysteresis and third quadrant presence) *In the context of using some kind of FTS samplers for Event-Based Control (EBC), the role of the hysteresis has been highlighted due to its capability of avoiding event generation due to noise [11].*

Besides, from the analysis perspective, the relative hysteresis h/δ *has been studied in works like [25], classifying non-linearities with hysteresis as "non-linearities with memory".*

Similarly to the conclusions reached in the aforementioned work, FTS samplers with hysteresis present a non-null imaginary part on the inverse negative of their DF. Conversely, "memoryless" FTS only have non-null real part.

In Fig. 3.6 three different FTS defined by $\Delta\epsilon/\epsilon{=}1/5$, $\Delta\delta/\delta{=}1/2$ *and* $\epsilon/\delta{=}0.75$ *have been presented, each with a different value for the relative hysteresis. In this figure it can be appreciated that the traces of the inverse negative of the DF tend to gradually fold towards the real axis with the reduction of the ratio* h/δ.

Property 3.4 (Bias influence in sampler's non-linear behavior) *By considering the generalized FTS, the induction of biased limit cycles is possible, i.e. limit cycle oscillations with the form* $x(t){=}A\sin(\omega t){+}B$ *with* $B \neq 0$ *at the input of the sampler.*

Regarding to the analytic study, the presence of bias can significantly change the shape of the inverse negative of the DF for a given FTS. For instance, consider any FTS that contains a deadband (some examples could be SSOD or RQH). If a biased oscillation is maintained with $B = \Delta\delta$, *then the behavior of these FTS will not correspond to a deadband, instead, the behavior will be similar to a multilevel relay, a sampler with no deadband. In that presented case, the bias B compensates the asymmetry* $\Delta\delta$, *changing completely the expected behavior.*

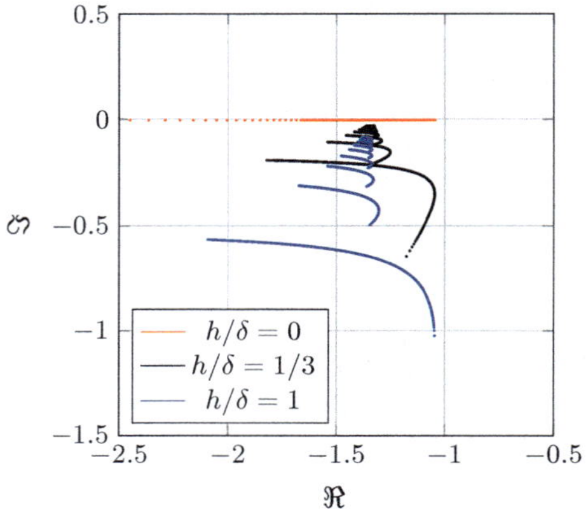

Fig. 3.6 Effect of the hysteresis on $-1/\mathcal{N}_A$ (Parameters used: $\Delta\epsilon/\epsilon = 1/5$, $\Delta\delta/\delta = 1/2$ and $\epsilon/\delta = 0.75$). (Reproduced from IEEE Access, Vol 10, Miguel-Escrig, O., Romero-Pérez, J. A., Sánchez-Moreno, J., Dormido, S., *Characterization of limit cycle oscillations induced by fixed threshold samplers*, ©IEEE 2022)

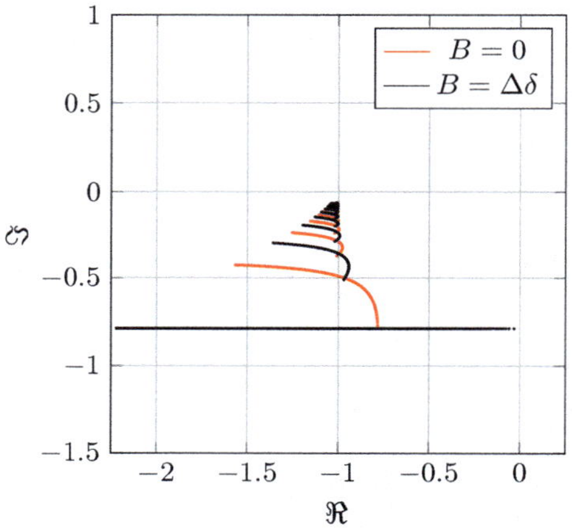

Fig. 3.7 Asymmetry compensation with the bias B and its effect on $-1/\mathcal{N}_A$ (Parameters used: $\Delta\epsilon/\epsilon = 1/2$, $\Delta\delta/\delta = 1/2$ and $\epsilon = \delta = h$). (Reproduced from IEEE Access, Vol 10, Miguel-Escrig, O., Romero-Pérez, J. A., Sánchez-Moreno, J., Dormido, S., *Characterization of limit cycle oscillations induced by fixed threshold samplers*, ©IEEE 2022)

From the analysis point of view, this compensation of the non-linear effect through the bias, implies that for any FTS in which $B = \Delta\delta$, a straight line that reaches the imaginary axis will appear on the Nyquist diagram. This line corresponds to the case of non-linearities without deadband and centered on the detection asymmetry (for visualization purposes, the produced effect results in a behavior similar to a relay).

In Fig. 3.7 a SSOD sampler has been presented illustrating two cases, one with null bias an another in which the bias compensates the detection asymmetry. In the last case, the horizontal line can be appreciated despite the fact that a SSOD sampler has been used.

Property 3.5 ($-1/\mathcal{N}_A$ periodicity with bias multiples) *Studying biased oscillations for a given sampler results in a different behavior depending on the bias. This is specifically reflected in the number of crossed levels in each direction, obtained from expressions in (3.5). For instance, starting from different values of B, the sequence of* *threshold crossings, resulting in the number of levels crossed in each direction varies when sweeping A. Therefore, the observed effect of having a biased oscillations is equivalent to have an asymmetry on the sampler.*

This effect can potentially hinder analysis study since the traces of $-1/\mathcal{N}_A$ *vary significantly with B. However, the placement of these traces has been observed to be repeated periodically in the Nyquist plane with the change of B. More specifically, the same placement of these traces is expected for:*

$$\frac{-1}{\mathcal{N}_A(B)} = \frac{-1}{\mathcal{N}_A(B')},$$

where $B' = B + k\delta$ *and* $k \in \mathbb{Z}$. *Observing the input-output relationship graph of a general FTS presented in Fig. 3.2, this property can be deduced since periodical shifts of bias results in a locally similar non-linearity. The same result can be proven analytically from the provided expressions.*

Property 3.6 ($-1/\mathcal{N}_A$ symmetry with regard to $\delta \times \epsilon$ square) *Another interesting property has been found which facilitates the analysis. Due to the characteristics of the FTS, it has been found that the traces of* $-1/\mathcal{N}_A$ *are placed in the spot for variations of B lesser than* δ. *In particular, it has been found the same placement for values of B symmetric with regard to the center and the extreme of* $\delta \times \epsilon$ *square.*

Synthetically, the placement of the DF traces will be the same for $B = \Delta\delta - \delta/2 + \alpha$ *and for* $B' = \Delta\delta - \delta/2 - \alpha$, $\alpha \in \Re$. *And also, with regard to the other symmetry axis, the same traces will be obtained with* $B = \Delta\delta + \beta$ *and* $B' = \Delta\delta - \beta$, $\beta \in \Re$.

Regarding to $\mathcal{N}_B$ it has not been found any new property that facilitates the analysis. Causes for this are that its application heavily depends on the elements on the loop. This fact does not hinder the analysis study using DIDF, which will be addressed in the following section.

3.4 Oscillation Conditions Study

The objective of this study is to provide guidelines on how to proceed with the analysis of this kind of systems. Therefore, it is not pursued to avoid or induct biased limit cycle oscillations, but to asses their appearance. From this guidelines further works are possible, for instance, conducting an analysis study to avoid limit cycle oscillations with FTS or to induct specific oscillations on the loop.

The first step towards assessing biased limit cycles sustainability is to determine if the bias is not attenuated through the loop and its value. The exact value is hardly ever known a priori, nonetheless, an initial guess can be done. Assuming reference or disturbance changes are high enough with regard to the detection threshold δ, the effect of the sampler can be substituted by a gain ϵ/δ, corresponding to the slope of the input-output relationship of the sampler (see Fig. 3.2). The initial guess of the bias, which will be referred to as bias central value B_c, can be estimated from expression (3.2) which results in:

$$B_c = \frac{r(\infty) - p(\infty)G(0)}{1 + \epsilon/\delta G_{ol}(0)}. \tag{3.6}$$

Due to the symmetry and multiplicity properties previously presented, a set of candidate biases is defined as $B_s \in [B_c - \delta/2, B_c + \delta/2]$.

Depending on the elements on the loop, the corresponding expression according to Table B.1 is chosen. Using the set B_s, the pairs $\{A, B\}$ fulfilling the given expression are obtained, which are the candidates to present limit cycle oscillations.

This pairs determine under which oscillation characteristics the bias is propagated, therefore, it must be evaluated the maintenance fo the oscillatory part of the signal. To assess this maintenance, the expression in (3.1) is evaluated for the given pairs. This produces a set of points in the Nyquist plane. If the open-loop transfer function intersects any of these points, that is:

$$G_{ol}(j\omega_o) = -\frac{1}{\mathcal{N}_A(A, B)},$$

then a limit cycle oscillation with $A\sin(\omega_o t) + B$ can be observed.

The following example illustrates this analysis methodology for several cases.

Example 3.1 Consider a process defined by the transfer function:

$$G(s) = \frac{1}{(s+1)^3},$$

also consider a sampler in the loop with $\Delta\epsilon/\epsilon = 0$, $\Delta\delta/\delta = 0$, $\epsilon/\delta = 1$ and $h/\delta = 1/2$. The input-output relationship of this sampler is shown in Fig. 3.8. No controller is placed in the loop ($C(s) = 1$).

According to Table B.1, this is a case where controller and process do not present poles nor zeros at the origin. Using the nomenclature in the table, $L_1(0) = 1$ and $L_2(0) = G(0)$, which results in the sustainability condition:

$$B(1 + \mathcal{N}_B G(0)) = r(\infty) - p(\infty)G(0). \tag{3.7}$$

It will be considered five scenarios:

1. Unitary reference change without disturbance.
2. Reference change of magnitude 0.95 without disturbance.
3. Unitary reference change and disturbance change of magnitude 0.95.
4. Same as scenario 3 but the gain of the process is increased.
5. Same as scenario 3 but the value ϵ of the sampler is doubled.

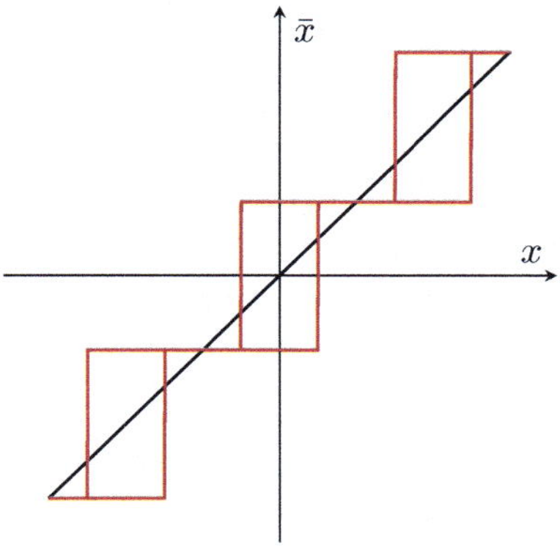

Fig. 3.8 Input-Output relationship of the sampler used in Example 3.1 (Parameters: $\Delta\epsilon/\epsilon = \Delta\delta/\delta = 0$, $\epsilon/\delta = 1$ and $h/\delta = 1/2$). (Reproduced from IEEE Access, Vol 10, Miguel-Escrig, O., Romero-Pérez, J. A., Sánchez-Moreno, J., Dormido, S., *Characterization of limit cycle oscillations induced by fixed threshold samplers*, ©IEEE 2022)

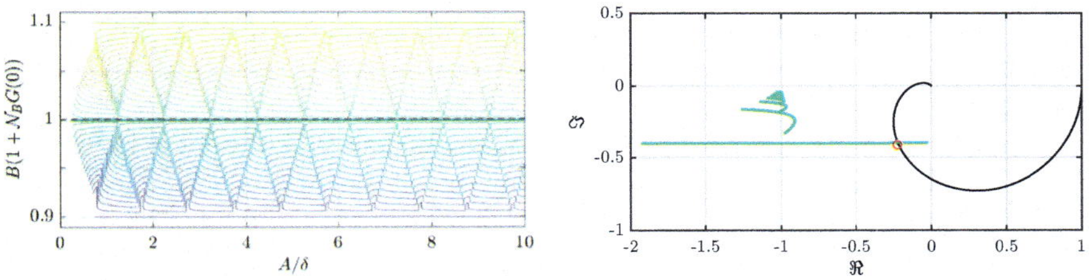

(a) Bias. (b) Dynamic.

Fig. 3.9 Graphical representation of the oscillation conditions for Scenario 1. (Reproduced from IEEE Access, Vol 10, Miguel-Escrig, O., Romero-Pérez, J. A., Sánchez-Moreno, J., Dormido, S., *Characterization of limit cycle oscillations induced by fixed threshold samplers*, ©IEEE 2022)

Scenario 1: The expression for the maintenance of the bias in this case results in:

$$B(1 + \mathcal{N}_B G(0)) = r(\infty). \tag{3.8}$$

The central value B_c is obtained from Eq. (3.6):

$$B_c = \frac{r(\infty)}{1 + \epsilon/\delta G(0)} = \frac{1}{2}.$$

Therefore, the set of biases to evaluate is defined as $B_s \in [B_c - \delta/2, B_c + \delta/2]$.

This set of biases is evaluated with expression (3.8), obtaining the corresponding pairs $\{A, B\}$. Graphically, since the left hand term changes with B, and the right hand term is constant, the values of A can be found in the intersection between both graphs.

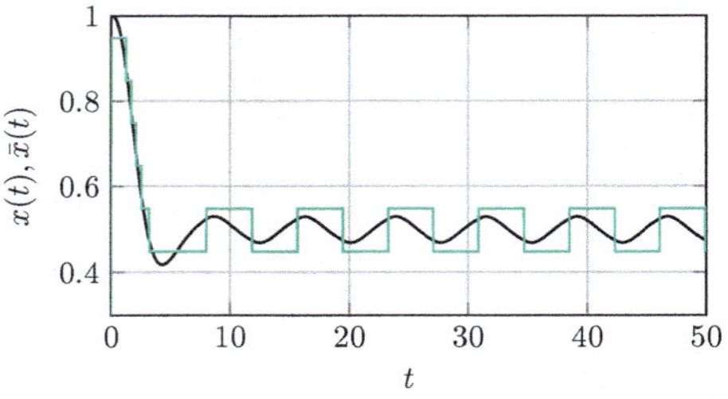

Fig. 3.10 Temporal response of the error $x(t)$ and sampled error $\bar{x}(t)$ signals under the conditions of the Scenario 1. (Reproduced from IEEE Access, Vol 10, Miguel-Escrig, O., Romero-Pérez, J. A., Sánchez-Moreno, J., Dormido, S., *Characterization of limit cycle oscillations induced by fixed threshold samplers*, ©IEEE 2022)

This is depicted on Fig. 3.9a, where each trace represents a bias within the set. From this figure, there is only one intersection between $r(\infty)$, the right hand term, presented in black, and the traces, but it coincides for all values of A. The trace that coincides corresponds with $B = 0.5$, and since all the values of A are solution, any intersection of the inverse negative of $\mathcal{N}_A$ for that value of B with the open-loop transfer function will produce a limit cycle oscillation.

The condition regarding to the oscillation maintenance is presented in Fig. 3.9b where it can be observed an intersection between $-1/\mathcal{N}(A, B = 0.5)$ and $G(j\omega)$.

The temporal response for this case has been obtained in simulation. The resulting error and sampled error signals are presented in Fig. 3.10. The oscillation frequency ω_o has been measured from this temporal response, and $G(j\omega_o)$ has been presented in Fig. 3.9b with a red circle, demonstrating the validity of the proposed analysis method.

Scenario 2: In this case, the expression (3.8) remains valid to evaluate the bias maintenance, the only difference being the magnitude of the change.

It has been followed the same procedure to obtain the set of biases B_s. Firstly, the central value of the bias has been obtained, resulting in $B_c = 0.475$, and then to obtain the set a variation around this value of $\pm\delta/2$ is established.

The set of biases have been used to evaluate Eq. (3.8), resulting in the representation presented in Fig. 3.11a, for its left hand term, where each trace, presented with a different color, corresponds to an item of the set. The dashed black line, which is the right hand term of that expression, has also been presented.

The pairs $\{A, B\}$ are obtained from the observed intersections, constituting the cases where the bias is maintained.

To assess the oscillation maintenance, in Fig. 3.11b the inverse negative of $\mathcal{N}_A$ for each value of B in the pair is presented in gray. Within these traces, the corresponding value of A in the pairs is highlighted, i.e. the specific points $-1/\mathcal{N}_A(A, B)$, where the pairs $\{A, B\}$ are solution of Eq. (3.8).

Note that the gray plots are not necessary for the analysis, but they have been included for illustrative purposes since they correspond to the classical DF analysis.

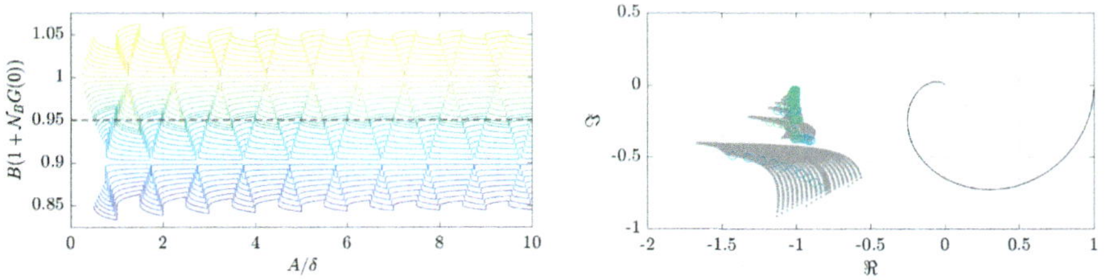

(a) Bias. (b) Dynamic.

Fig. 3.11 Graphical representation of the oscillation conditions for Scenario 2. (Reproduced from IEEE Access, Vol 10, Miguel-Escrig, O., Romero-Pérez, J. A., Sánchez-Moreno, J., Dormido, S., *Characterization of limit cycle oscillations induced by fixed threshold samplers*, ©IEEE 2022)

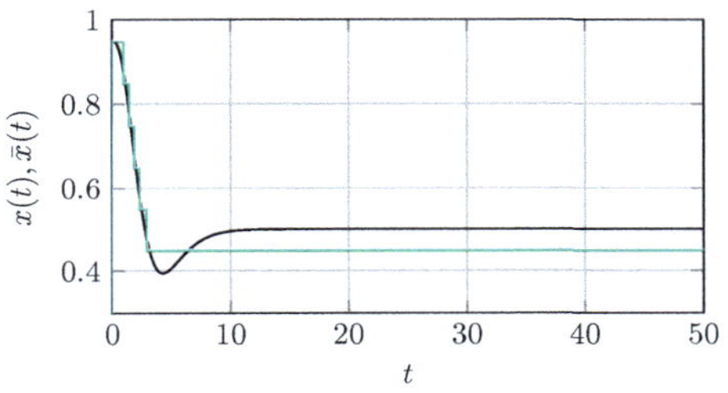

Fig. 3.12 Temporal response of the error $x(t)$ and sampled error $\bar{x}(t)$ signals under the conditions of the Scenario 2

Regarding to the graphical depiction, the open-loop transfer function does not present an intersection with the highlighted points, therefore, no limit cycle oscillations take place. This has been corroborated in simulation. As in the precedent case, error and sampled error signals are depicted in Fig. 3.12.

Scenario 3: Since the signals present in this scenario are, not only the reference change, but also the disturbance change, the expression to estimate the central value B_c is deduced from Eq. (3.6) is:

$$B_c = \frac{r(\infty) - p(\infty)G(0)}{1 + \epsilon/\delta G(0)} = \frac{1 - 0.95}{2} = 0.025.$$

As in precedent scenarios, a set of biases B_s has been obtained, and, it has been used to evaluate the expression (3.7), whose left hand term is represented in Fig. 3.13a. The right hand term of that equation has also been presented with a dashed black line. The candidate set of pairs $\{A, B\}$ is obtained from the intersections observed in Fig. 3.13a. As in the previous scenario, the dynamic part of the DIDF has been evaluated, and, in Fig. 3.13b, no intersection is observed between the open-loop transfer function and the points defined by $-1/\mathcal{N}_A(A, B)$. Thus, no limit cycle oscillation is induced in the loop.

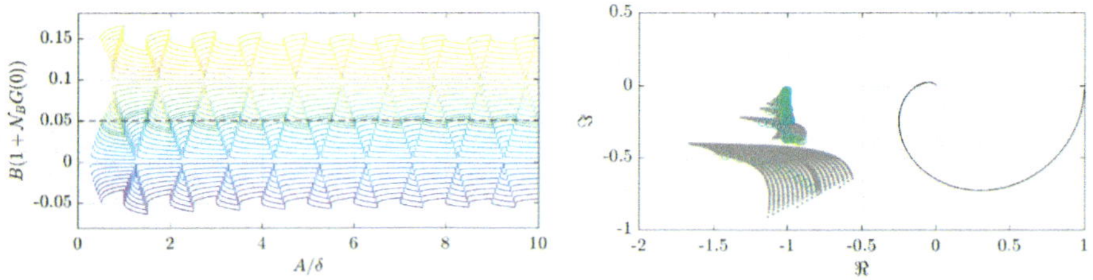

(a) Bias. (b) Dynamic.

Fig. 3.13 Graphical representation of the oscillation conditions for Scenario 3. (Reproduced from IEEE Access, Vol 10, Miguel-Escrig, O., Romero-Pérez, J. A., Sánchez-Moreno, J., Dormido, S., *Characterization of limit cycle oscillations induced by fixed threshold samplers*, ©IEEE 2022)

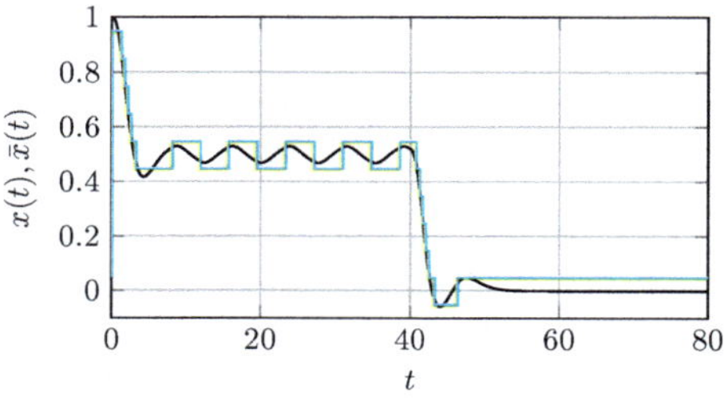

Fig. 3.14 Temporal response of the error $x(t)$ and sampled error $\bar{x}(t)$ signals under the conditions of the Scenario 3

This scenario shows that, in this type of loop configuration, disturbance input changes can stabilize or destabilize the temporal response. This effect can be seen in Fig. 3.14, where the simulated response for this scenario is presented. In the first half of the simulation, only the reference change is applied, corresponding to Scenario 1, where the oscillatory behavior was proven. Nevertheless, in this scenario also the influence of a disturbance change is studied, which is applied at $t = 40\ s$, and as predicted in the analysis, the process stabilizes in the surroundings of the central bias.

Scenario 4: In this scenario reference and disturbance changes are evaluated, and, in addition, a process gain 4 times bigger than the original is considered. To avoid repetition, the steps that repeat from precedent scenarios are omitted.

The oscillation conditions graphical representation is depicted in Fig. 3.15. As it can be seen from this graphical representation, it exist an intersection in the Nyquist pane between a candidate point belonging to the inverse negative of the dynamic describing function and the open-loop transfer function. This reveals the existence of a limit cycle oscillation.

This system has been tested in simulation obtaining the temporal response depicted in Fig. 3.16. In the presented experience, the reference change is applied at $t = 0\ s$ and the disturbance change at $t = 40\ s$.

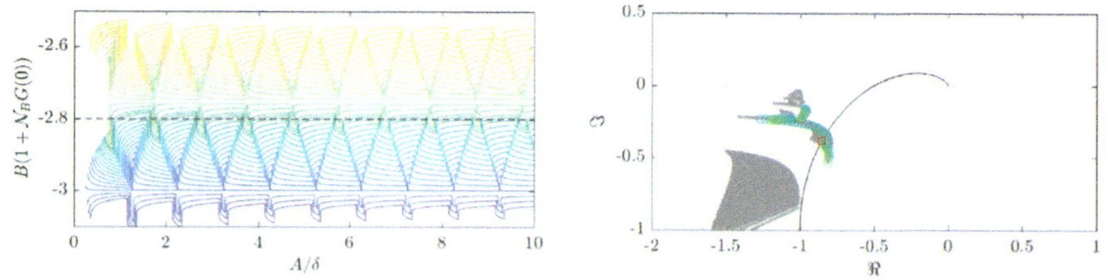

(a) Bias. (b) Dynamic.

Fig. 3.15 Graphical representation of the oscillation conditions for Scenario 4. (Reproduced from IEEE Access, Vol 10, Miguel-Escrig, O., Romero-Pérez, J. A., Sánchez-Moreno, J., Dormido, S., *Characterization of limit cycle oscillations induced by fixed threshold samplers*, ©IEEE 2022)

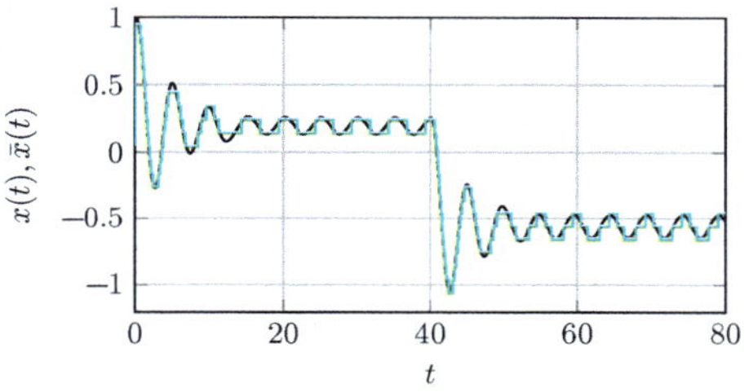

Fig. 3.16 Temporal response of the error $x(t)$ and sampled error $\bar{x}(t)$ signals under the conditions of the Scenario 4

This scenario illustrates how two different limit cycles can be obtained for the same sampler depending on the signal excitation. The first oscillation, resulting from the step change, is characterized by $m_{sup} + m_{inf} = 1$, and, while it has not been evaluated theoretically, the procedure is the same that in Scenario 1. The second oscillation corresponds to the evaluated case, with $m_{sup} = m_{inf} = 1$. The oscillation frequency has been measured and $G(j\omega_o)$ highlighted with a red square in Fig. 3.15a, demonstrating the validity of the approach.

This example illustrates an interesting case where two oscillatory modes are obtained using the same sampler. Applications of this fact could be related to identification field.

Scenario 5: For this scenario consider the ratio $\epsilon/\delta = 2$. This introduces a change in the calculation of the central bias:

$$B_c = \frac{r(\infty) - p(\infty)G(0)}{1 + \epsilon/\delta G(0)} = \frac{1 - 0.95}{3} = 0.0167.$$

Similarly to previous cases, the set of biases B_s is calculated and the maintenance of biased oscillations is evaluated. Graphically, the conditions are depicted in Fig. 3.17.

Regarding to the graphical representation, a limit cycle oscillation is forecast.

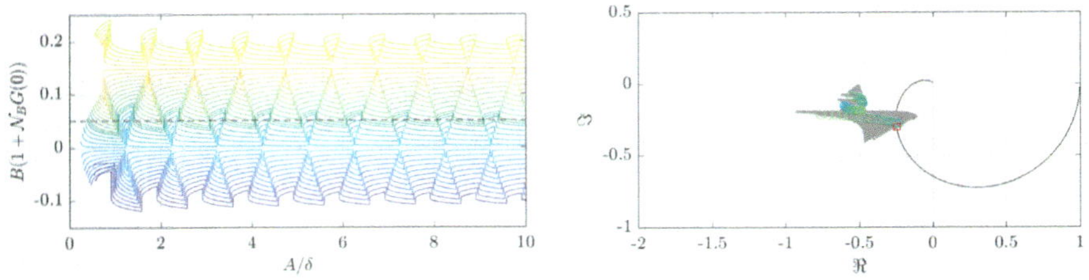

(a) Bias. (b) Dynamic.

Fig. 3.17 Graphical representation of the oscillation conditions for Scenario 5. (Reproduced from IEEE Access, Vol 10, Miguel-Escrig, O., Romero-Pérez, J. A., Sánchez-Moreno, J., Dormido, S., *Characterization of limit cycle oscillations induced by fixed threshold samplers*, ©IEEE 2022)

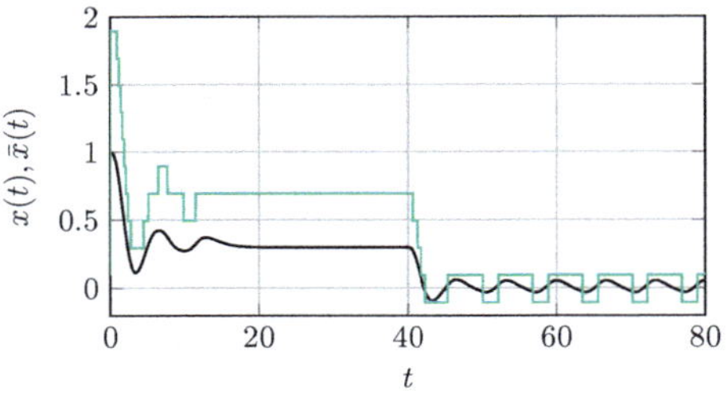

Fig. 3.18 Temporal response of the error $x(t)$ and sampled error $\bar{x}(t)$ signals under the conditions of the Scenario 5. (Reproduced from IEEE Access, Vol 10, Miguel-Escrig, O., Romero-Pérez, J. A., Sánchez-Moreno, J., Dormido, S., *Characterization of limit cycle oscillations induced by fixed threshold samplers*, ©IEEE 2022)

The temporal response of the system is presented in Fig. 3.18. The oscillation frequency has been measured, and, $G(j\omega_o)$ has been marked in a red square in the Nyquist diagram in Fig. 3.17b, validating the analysis.

This scenario illustrates a case where, starting from a stable response, the excitation signal induces a limit cycle in the temporal response.

Quick summary:

- ☐ Multilevel Fixed Threshold Samplers (FTS) have been parameterized, gathering the behavior of other well-known quantifiers in a generalized manner.
- ☐ Tools for characterizing limit cycle oscillations induced by FTS have been provided.
- ☐ The Dual Input Describing Function (DIDF) is employed to analyze the sustainability of biased oscillations induced by FTS.

☐ The study of the DIDF for FTS reveals a series of properties, such as periodicity or symmetry, that facilitate the analysis.
☐ Guidelines for robustness analysis are provided, emphasizing the significant impact of the loop configuration and involved signals on oscillation conditions.
☐ Illustrative examples applying the analysis methodology are provided, showcasing different scenarios found in this kind of systems.

References

1. Samad T (2017) A survey on industry impact and challenges thereof. IEEE Control Syst 37(1):17–18
2. Maxim A, Copot D, Copot C, Ionescu CM (2019) The 5w's for control as part of industry 4.0: why, what, where, who, and when-A PID and MPC control perspective. Inventions 4(1)
3. Årzén K-E (1999) A simple event-based PID controller. In: Proceedings of the 14th IFAC world congress, vol 18, pp 423–428
4. Durand S, Marchand N (2009) An event-based PID controller with low computational cost. In: 8th international conference on sampling theory and applications (SampTA'09), pages Special–session
5. Durand S, Marchand N (2009) Further results on event-based PID controller. In: Control conference (ECC), 2009 European. IEEE, pp 1979–1984
6. Vasyutynskyy V, Kabitzsch K (2009) Time constraints in PID controls with send-on-delta. IFAC Proc Vol 42(3):48–55
7. Vasyutynskyy V, Kabitzsch K (2010) A comparative study of PID control algorithms adapted to send-on-delta sampling. In: 2010 IEEE International Symposium on Industrial Electronics (ISIE). IEEE, pp 3373–3379
8. Miguel-Escrig O, Romero-Pérez J-A (2018) Implementation and evaluation of event-based PID in the IEC-61499 standard. In: 2018 IEEE 23rd international conference on emerging technologies and factory automation (ETFA), vol 1, pp 574–580. IEEE
9. Zoitl A, Lewis R (2014) Modelling control systems using IEC 61499, volume 95. IET
10. Beschi M, Dormido S, Sánchez J, Visioli A (2012) Characterization of symmetric send-on-delta PI controllers. J Process Control 22(10):1930–1945
11. Miguel-Escrig O, Romero-Pérez J-A (2020) Regular quantisation with hysteresis: a new sampling strategy for event-based PID control systems. IET Control Theory & Applications
12. Sánchez J, Guinaldo M, Visioli A, Dormido S (2018) Identification of process transfer function parameters in event-based PI control loops. ISA Trans 75:157–171
13. Sánchez J, Guinaldo M, Visioli A, Dormido S (2018) Enhanced event-based identification procedure for process control. Ind Eng Chem Res 57(21):7218–7231
14. Sánchez J, Guinaldo M, Visioli A, Dormido S (2019) Identification and tuning methods for PI control systems based on symmetric send-on-delta sampling. Int J Control Autom Syst 17(11):2784–2795
15. Beschi M, Dormido S, Moreno JS, Visioli A, Yebra LJ (2014) Event-based PI plus feedforward control strategies for a distributed solar collector field. IEEE Trans Contr Sys Tech 22:1615–1622
16. Pawlowski A, Beschi M, Guzmán JL, Visioli A, Berenguel M, Dormido S (2016) Application of SSOD-PI and PI-SSOD event-based controllers to greenhouse climatic control. ISA Trans 65:525–536

17. Romero-Pérez J-A, Sanchis-Llopis R, Arrebola E (2015) Experimental study of event based PID controllers with different sampling strategies. Application to brushless DC motor networked control system. In: 2015 XXV international conference on information, communication and automation technologies (ICAT), pp 1–6, Oct 2015
18. Beschi M, Dormido S, Sánchez J, Visioli A (2014) Tuning of symmetric send-on-delta proportional-integral controllers. IET Control Theory Appl 8:248–259(11)
19. Romero-Pérez J-A, Sanchis-Llopis R (2016) A new method for tuning PI controllers with symmetric send-on-delta sampling strategy. ISA Trans 64:161–173
20. Miguel-Escrig O, Romero-Pérez J-A, Sanchis-Llopis R (2020) Tuning PID controllers with symmetric send-on-delta sampling strategy. J Frank Inst 357(2):832–862
21. Sánchez J, Guinaldo M, Dormido S, Visioli A (2020) Validity of continuous tuning rules in event-based PI controllers using symmetric send-on-delta sampling: an experimental approach. Comput Chem Eng 106878
22. West JC, Douce JL, Livesley RK (1956) The dual-input describing function and its use in the analysis of non-linear feedback systems. Proc IEE-Part B: Radio Electron Eng 103(10):463–473
23. Romero-Pérez J-A, Sanchis-Llopis R, Peñarrocha-Alós I (2014) A simple rule for tuning event-based PID controllers with symmetric send-on-delta sampling strategy. In: Proceedings of the 2014 IEEE emerging technology and factory automation (ETFA), , Sept 2014, pp 1–8
24. Romero-Pérez J-A, Sanchis-Llopis R (2017) Tuning and robustness analysis of event-based PID controllers under different event generation strategies. Int J Control 1–38
25. Gelb A, Van der Velde WE (1968) Multiple-input describing functions and non-linear system design. McGraw-Hill

Chapter 4
Analysis of SSOD and RQH Systems with Tsypkin Method

4.1 Introduction

Presently, Event-Based Control (EBC) emerges as a promising alternative to traditional time-driven control systems, aiming to diminish the required measurement frequency for control without compromising closed-loop performance. This proves essential for controllers in networked control systems, where numerous devices (sensors, actuators, controllers) share a limited bandwidth communication channel. The reduction in transmitted messages not only enhances the ov erall behavior of the network, preventing issues like dropouts and delays, but also aligns with the growing utilization of wireless communications in control applications. In such scenarios, minimized data transmission leads to a significant reduction in power consumption, thereby prolonging the battery life of self-powered remote sensors [1].

In Event-Based Control Systems (EBCS), the measurement frequency is significantly influenced by the strategy employed to generate events. One promising approach in this regard is the Send-On-Delta (SOD) strategy, which effectively minimizes the transmission of new data [2, 3]. The SOD method involves transmitting data from the sensor to the controller node only when the measured value exceeds a specified threshold δ [4]. A specific variant of SOD, known as Symmetric-Send-On-Delta (SSOD), is introduced in [5]. The SSOD sampler quantifies the input signal by a multiple of a constant value δ, and then transmits a new value when the input changes by δ units. A variant on this sampler, called Regular Quantization with Hysteresis (RQH), has been recently proposed in [6], proving to improve efficiency in event generation.

One crucial aspect in the analysis and design of EBCS is the presence of limit cycles, which lead to sustained oscillations in the closed-loop response. The characterization of limit cycles for various types of process models, including Integrator processes Plus Time Delay (IPTD), First Order processes Plus Time Delay (FOPTD), and Second Order processes Plus Time Delay (SOPTD), using a SSOD sampling strategy has been outlined in [7]. Understanding limit cycle appearance is essential

O. Miguel-Escrig, *Event-Based PID Controllers with Fixed Threshold Sampling Strategies*, Springer Theses, https://doi.org/10.1007/978-3-031-62505-3_4

to forecast the behavior of a system due to dynamic changes in the process. From a PID tuning perspective, a primary objective is to prevent limit cycle oscillations that could diminish the overall performance of the control system or cause excessive actuator wear.

Several works have addressed the limit cycle oscillation issue in SSOD loops under the perspective of the Describing Function (DF) technique [8–10]. This method is well-known for its ease of use and allows applying concepts from the classical continuous control domain to EBC. Among these concepts, the Nyquist diagram and subsequent frequency analysis can be found.

DF applicability relies on the filtering characteristics of the open-loop transfer function, which is responsible of attenuating higher order harmonics to allow the successful application of the DF. Consequently, low-order models, including FOPTD and SOPTD, commonly employed to depict real-world industrial processes, cannot be studied with this approach. Besides, the introduction of the derivative term in the controller, also hinders the attenuation of such harmonics. Thus, the development of alternative methods that encompass a wider spectrum of processes and controllers is necessary.

In this chapter, a novel robustness measure aimed at preventing limit cycles in SSOD based control systems is presented. In contrast to those presented in previous works like [11], the proposed robustness margins are applicable to low-order systems. The approach is grounded in Tsypkin's method [12], extensively employed in the analysis of relay control systems. As a result, a new margin M_T called Tsypkin's margin is obtained, and it is proven to be sufficient to avoid also multi-leveled limit cycle oscillations. The newly derived margin is employed to evaluate the robustness against limit cycle oscillations in SSOD-PID control systems, particularly when continuous-time tuning methods are employed. Specifically, Ziegler-Nichols [13]), Cohen-Coon [14], and AMIGO [15] tuning methods for controlling FOPTD processes in SSOD-$C(s)$ loops are evaluated. A study aimed to extend the applicability of the proposed margin to loops with RQH sampling schemes is also presented.

4.2 Problem Statement

The networked control system considered in this chapter is presented in Fig. 2.1. The blocks $C(s)$ and $G_p(s)$ correspond to the controller and process transfer functions, respectively, the ZOH block is a zero-order hold, the network is modeled by a delay t_d and a SSOD sampling strategy is performed at the sensor unit in the block EG. The signals presented in the figure are the reference signal y_r, controlled output y, error e and sampled and held error $\bar{e}$.

This scheme was proposed in [5], and together with contributions in [16], presented some tuning rules developed for PI controllers. The type of processes and controllers in those works facilitate the applicability of analysis techniques like the Describing Function (DF), applied in works like [8, 9]. However, the filtering hypothesis of the DF is not fulfilled when considering low order models, like a First Order

Plus Time Delay (FOPTD), or adding a derivative term to the controller. For the sake of completeness, it will be considered the following form for the PID controller:

$$C(s) = K_p \left(1 + \frac{1}{T_i s} + \frac{N T_d s}{T_d s + N}\right).$$

To illustrate the limited validity of this kind of technique to forecast the induction of limit cycles the following example is introduced.

Example 4.1 Consider a process characterized by the following transfer function:

$$G(s) = \frac{e^{-0.2s}}{s+1}. \tag{4.1}$$

Several PID controllers have been tune according to Ziegler-Nichols [13], Cohen-Coon [14], AMIGO [15] and Sanchis-Romero-Balaguer [17] (SRB for the sake of brevity) methods. Their parameters are presented compactly in Table 4.1.

To assess the robustness for this kind of systems, the loop in Fig. 2.1 is transformed to the Hammerstein-Wiener representation in Fig. 4.1, where SSOD and ZOH blocks have been combined.

According to the DF theory, limit cycles oscillations are avoided if:

$$G_{ol}(j\omega) \neq -\frac{1}{\mathcal{N}}, \quad \forall \omega,$$

Table 4.1 PID Parameters with different tuning rules for the system presented in Eq. (4.1)

	$\mathbf{K_p}$	T_i	T_d	N
Ziegler-Nichols	6	0.4	0.1	10
Cohen-Coon	6.917	0.455	0.07	10
AMIGO	2.45	0.587	0.094	10
SRB	2.181	0.484	0.115	10

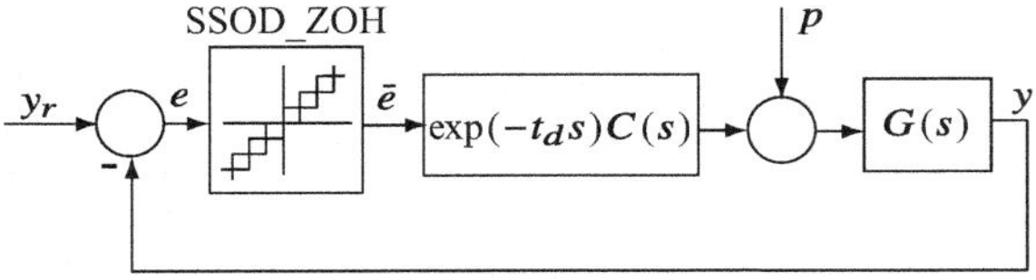

Fig. 4.1 Non-linear equivalent system to the control system with SSOD sampling strategy in Fig. 2.1. (Reprinted from IFAC-PapersOnLine, Vol 51, No. 4, Miguel-Escrig, O., Romero-Pérez, J. A., & Sanchis-Llopis, R., *New robustness measure for a kind of event-based PID*, 781–786, Copyright (2018), with permission from Elsevier)

where $G_{ol}(j\omega)$ is the open-loop transfer function:

$$G_{ol}(j\omega) = \exp(-t_d j\omega)C(j\omega)G(j\omega)$$

and $\mathcal{N}$ the describing function corresponding to the SSOD_ZOH block.

The oscillation condition indicates the induction of limit cycle oscillations is possible if it exists an intersection between the open-loop transfer function and the traces of $-1/\mathcal{N}$ at any frequency.

This oscillation condition has been evaluated for the process and controllers previously presented and the $-1/\mathcal{N}$ traces of the SSOD. An intersection is detected only for the systems controlled with Ziegler-Nichols and Cohen-Coon methods, for which is expected the induction of a sinusoidal limit cycle oscillation.

The temporal response of these systems has been obtained in simulation. These responses offer hardly sinusoidal limit cycle oscillations in the four cases, contradicting the analysis performed using the DF, which studies only sinusoidal oscillations.

This example evinces the limited validity of DF to forecast the behavior of systems, characterized by their open-loop transfer function, with low filtering capabilities.

4.3 New Robustness Margin Against Limit Cycles Induction: The Tsypkin's Method

Previous section showed the limitations of the DF in forecasting the existence of limit cycle oscillations in loops like the presented in Fig. 2.1. Therefore, in this section an approach based on Tsypkin's method [12] is presented. This method remains accurate regardless of the filtering capabilities of the linear part of the system, thus, allowing its applicability to FOPTD systems and PID control.

Consider the system in the condensed form presented in Fig. 4.1. A generic limit cycle oscillation sampled by the non-linear block SSOD_ZOH corresponds to a sampled error signal $\bar{e}$ with the form of a stair-like signal, in which the step widths are given by $(\rho_p - \rho_{p-1})T_o/2$, where T_o is the oscillation period and $\rho_i \in]0, 1[$. Note that the specific form of error signal e is not relevant, only the slopes at the commutation time fractions $\rho_i T_o/2$.

This generic depiction is mathematically formalized in the following set of conditions:

$$EC_p \left| e\left(\rho_p \frac{T_o}{2}\right) = \begin{cases} p\delta & \text{for } p = 1, 2...m \\ (2m - p)\delta & \text{for } p = m+1, ...2m \end{cases} \right. \tag{4.2}$$

$$IC_p \left| \frac{de\left(\rho_p \frac{T_o}{2}\right)}{dt} = \begin{cases} > 0 & \text{for } p = 1, 2...m \\ < 0 & \text{for } p = m+1, ...2m \end{cases} \right. \tag{4.3}$$

Both expressions are referred to the error signal, conditions in (4.2) correspond to the level crossing and expressions in (4.3) to the derivative sign.

A general expression for $e(\rho_p \frac{T_o}{2})$ can be obtained through Fourier series analysis (see Appendix A), resulting in Eq. (4.4), where n is the number of harmonics.

$$
\begin{aligned}
e\left(\rho_p \frac{T_o}{2}\right) = &-\frac{4\delta}{\pi} \sum_{n=1,3,5,\ldots}^{\infty} \frac{1}{n} \Re\{G_{ol}(jn\omega_o)\} \left(\sum_{i=1}^{m} i \sin\left(\frac{n\pi}{2}(\rho_{i+1}-\rho_i)\right) \cos\left(n\pi\left(\frac{\rho_{i+1}+\rho_i}{2} - \rho_p\right)\right) \right. \\
&\left. + \sum_{i=m+1}^{2m-1} (2m-i) \sin\left(\frac{n\pi}{2}(\rho_{i+1}-\rho_i)\right) \cos\left(n\pi\left(\frac{\rho_{i+1}+\rho_i}{2} - \rho_p\right)\right) \right) \\
&-\frac{4\delta}{\pi} \sum_{n=1,3,5,\ldots}^{\infty} \frac{1}{n} \Im\{G_{ol}(jn\omega_o)\} \left(\sum_{i=1}^{m} i \sin\left(\frac{n\pi}{2}(\rho_{i+1}-\rho_i)\right) \sin\left(n\pi\left(\frac{\rho_{i+1}+\rho_i}{2} - \rho_p\right)\right) \right. \\
&\left. + \sum_{i=m+1}^{2m-1} (2m-i) \sin\left(\frac{n\pi}{2}(\rho_{i+1}-\rho_i)\right) \sin\left(n\pi\left(\frac{\rho_{i+1}+\rho_i}{2} - \rho_p\right)\right) \right)
\end{aligned}
\tag{4.4}
$$

The obtained expression is a function of real and imaginary part of the open-loop transfer function ($\Re\{G_{ol}(jn\omega_o)\}$, $\Im\{G_{ol}(jn\omega_o)\}$), harmonics ($n$), and number of thresholds crossed (m).

Analysis with this generic expressions becomes unmanageable when considering $m > 1$ due to the number of conditions in (4.2) and (4.3) to satisfy and the resulting expression in (4.4). At this point, to continue with the analytic procedure, Conjecture 4.1 is assumed.

Conjecture 4.1 *For the systems under study, avoiding* $m = 1$ *limit cycle oscillations assures that* $m > 1$ *oscillations cannot take place.*

Reader is referred to Sect. 4.4 for further development.

Focusing on $m = 1$ oscillations, the conditions for limit cycle existence are simplified:

$$
EC_{p|m=1} \left| e\left(\rho_p \frac{T_o}{2}\right) = \begin{cases} \delta & \text{for } p = 1 \\ 0 & \text{for } p = 2 \end{cases} \right.
\tag{4.5}
$$

$$
IC_{p|m=1} \left| \frac{de\left(\rho_p \frac{T_o}{2}\right)}{dt} = \begin{cases} > 0 & \text{for } p = 1 \\ < 0 & \text{for } p = 2 \end{cases} \right. .
\tag{4.6}
$$

These conditions are sufficient and necessary for the existence of $m = 1$ limit cycle oscillations.

Conditions in (4.5) have proven to provide concrete results in limit cycle maintenance. Expression (4.4) is evaluated with $m = 1$ and $p = 1, 2$ and equated with (4.5). By definition $\rho_{2m} = 1$ and renaming ρ_1 simply as ρ, it is obtained:

$$
\begin{aligned}
\sum_{n=1,3,5,\ldots}^{\infty} \frac{1}{n} \Re\{G_{ol}(jn\omega_o)\} \sin(n\pi\rho) + \frac{\pi}{4} &= 0 \\
\sum_{n=1,3,5,\ldots}^{\infty} \frac{1}{n} \Im\{G_{ol}(jn\omega_o)\} \cos^2\left(\frac{n\pi\rho}{2}\right) + \frac{\pi}{8} &= 0
\end{aligned}
\tag{4.7}
$$

and isolating:

$$
\begin{aligned}
\Re\{G_{ol}(j\omega_o)\} &= -\frac{\frac{\pi}{4} + \sum\limits_{n=3,5..}^{\infty} \frac{1}{n} \sin(n\pi\rho)\, \Re\{G_{ol}(jn\omega_o)\}}{\sin(\pi\rho)}, \\
\Im\{G_{ol}(j\omega_o)\} &= -\frac{\frac{\pi}{8} + \sum\limits_{n=3,5..}^{\infty} \frac{1}{n} \cos^2\left(\frac{n\pi\rho}{2}\right) \Im\{G_{ol}(jn\omega_o)\}}{\cos^2\left(\frac{\pi\rho}{2}\right)}.
\end{aligned}
\tag{4.8}
$$

Definition 4.1 A Tsypkin branch $B_T(\omega, \rho)$ for SSOD_ZOH samplers is defined according to expression (4.9) for a given frequency ω and $\rho \in]0, 1[$ as the trajectory in the Nyquist diagram resulting from the right hand members of expressions in (4.8).

$$
B_T(\omega, \rho) = \left(-\frac{\frac{\pi}{4} + \sum\limits_{n=3,5..}^{\infty} \frac{1}{n} \sin(n\pi\rho)\, \Re\{G_{ol}(jn\omega)\}}{\sin(\pi\rho)}, \right.
$$
$$
\left. -\frac{\frac{\pi}{8} + \sum\limits_{n=3,5..}^{\infty} \frac{1}{n} \cos^2\left(\frac{n\pi\rho}{2}\right) \Im\{G_{ol}(jn\omega)\}}{\cos^2\left(\frac{\pi\rho}{2}\right)} \right), \quad \forall \rho \in]0, 1[\tag{4.9}
$$

Theorem 4.1 *A limit cycle oscillation with frequency ω_o, oscillation amplitude lesser than 2δ and commutation time fraction ρ_o will take place in SSOD-$C(s)$ loops if:*

$$
\exists \omega_o, \rho_o : (\Re\{G_{ol}(j\omega_o)\}, \Im\{G_{ol}(j\omega_o)\}) = B_T(\omega_o, \rho_o). \tag{4.10}
$$

where $G_{ol}(j\omega)$ is the open-loop transfer function and $B_T(\omega, \rho)$ a Tsypkin branch as in Definition 4.1.

Corollary 4.1 *The Tsypkin's robustness margin M_T for SSOD-$C(s)$ loops is defined as:*

$$
M_T(\omega_{min}, \rho_{min}) = \min_{\omega, \rho}\{dist(G_{ol}(j\omega), B_T(\omega, \rho))\} \tag{4.11}
$$

and represents the minimum euclidean distance between the open-loop transfer function and any Tsypkin branch. The analytic expression for M_T is presented in Eq. (4.12).

$$M_T(\omega_{min}, \rho_{min}) = \sqrt{\left(\frac{\frac{\pi}{4} + \sum_{n_{odd}}^{\infty} \frac{1}{n} \sin(n\pi\rho_{min}) \Re\{G_{ol}(jn\omega_{min})\}}{\sin(\pi\rho_{min})}\right)^2 + \left(\frac{\frac{\pi}{8} + \sum_{n_{odd}}^{\infty} \frac{1}{n} \cos^2\left(\frac{n\pi\rho_{min}}{2}\right) \Im\{G_{ol}(jn\omega_{min})\}}{\cos^2\left(\frac{\pi\rho_{min}}{2}\right)}\right)^2} \quad (4.12)$$

The application of the newly obtained oscillation condition and robustness measure is illustrated in the following examples.

Example 4.2 For the following FOPTD process model:

$$G(s) = \frac{e^{-s}}{s+1}.$$

A PI controller according to SRB method [17] is tuned (parameters: $K_p = 0.84$, $T_i = 1.17$).

Tsypkin branches B_T have been calculated according to Eq. (4.9) using frequencies $\omega \in [\frac{\omega_{cg}}{200}, \omega_{cg}]$, where ω_{cg} is the crossover frequency. Several of these branches have been plotted in the Nyquist diagram in Fig. 4.2 together with the open-loop transfer function.

Each branch and its corresponding point in the frequency response, i.e. the point $(\Re\{G_{ol}(j\omega_k)\}, \Im\{G_{ol}(j\omega_k)\})$ with same frequency that the branch $B_T(\omega_k)$, is presented with a different color. The Tsypkin branch that defines the robustness margin M_T $(B_T(\omega_{min}))$ is presented with a dashed black line.

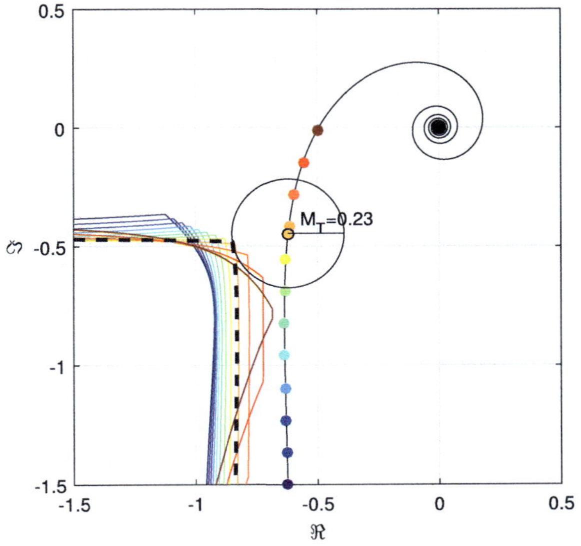

Fig. 4.2 Nyquist diagram of $G_{ol}(\omega)$ and the Tsypkin band for a non-oscillating system ($M_T = 0.23$). (Reprinted from Journal of the Franklin Institute, Vol 357, No. 2, Miguel-Escrig, O., Romero-Pérez, J. A., & Sanchis-Llopis, R., *Tuning PID controllers with symmetric send-on-delta sampling strategy*, 832–862, Copyright (2020), with permission from Elsevier)

The measured Tsypkin margin for this example is $M_T = 0.23$, obtained for $\omega = 1.0191 rad/s$. Since a positive margin is obtained, no limit cycle oscillations are expected, which has been corroborated in simulation.

Remark 4.1 Unlike in DF theory, considering uniquely the placement of the Tsypkin branches does not suffice to assess system's robustness. The placement of the point in the frequency response that matches in frequency must be considered. In precedent example, red traces (see Fig. 4.2) are placed closer to the open-loop transfer function, but they do not define M_T.

Example 4.3 The system presented in Example 4.1 is retaken. Specifically the SRB controller is analyzed. Similarly to the analysis undergone in precedent example, the Tsypkin branches have been calculated and an intersection has been detected, leading to $M_T = 0$, thus, corroborating the apparition of a limit cycle oscillation.

This process and controller configuration was demonstrated to oscillate in simulations conducted in Example 4.1, and proper analysis through Tsypkin margin leads to the same result. Concretely, the intersection between $G_{ol}(j\omega)$ and $B_T(\omega)$ can be found at $\omega = \omega_o = 1.55$ rad/s.

4.4 Multi-level Oscillations

In previous section oscillations with $m = 1$ have been characterized through Tsypkin method, resulting in the robustness measure M_T. The provided analysis was focused on $m = 1$ oscillations, based on Conjecture 4.1. In this section the study for $m > 1$ oscillations is addressed.

Calculus complexity for evaluation of Tsypkin's conditions escalates with m. For illustrative purposes, the case of $m = 2$ is presented, but the procedure holds for any value of m.

Considering, then, $m = 2$ results in the following oscillation conditions, (see Eq. (4.2)):

$$EC_{p|m=2} \left| e\left(\rho_p \frac{T_0}{2}\right) = \begin{cases} \delta & \text{for } p = 1 \\ 2\delta & \text{for } p = 2 \\ \delta & \text{for } p = 3 \\ 0 & \text{for } p = 4 \end{cases} \right., \tag{4.13}$$

and evaluating Eq. (4.4) for $m = 2$ results in expression (4.14).

$$
\begin{aligned}
e\left(\rho_p \frac{T_0}{2}\right) = -\frac{4\delta}{\pi}\sum_{n_{odd}}^{\infty}\frac{1}{n}\Bigg(\Re\{G_{ol}(jn\omega_0)\}\Bigg(&\sin\left(\frac{n\pi}{2}(\rho_2-\rho_1)\right)\cos\left(n\pi\left(\frac{\rho_2+\rho_1}{2}-\rho_p\right)\right)\\
&+2\sin\left(\frac{n\pi}{2}(\rho_3-\rho_2)\right)\cos\left(n\pi\left(\frac{\rho_3+\rho_2}{2}-\rho_p\right)\right)\\
&+\sin\left(\frac{n\pi}{2}(\rho_4-\rho_3)\right)\cos\left(n\pi\left(\frac{\rho_4+\rho_3}{2}-\rho_p\right)\right)\Bigg)\Bigg)\\
+\frac{4\delta}{\pi}\sum_{n_{odd}}^{\infty}\frac{1}{n}\Bigg(\Im\{G_{ol}(jn\omega_0)\}\Bigg(&\sin\left(\frac{n\pi}{2}(\rho_2-\rho_1)\right)\sin\left(n\pi\left(\frac{\rho_2+\rho_1}{2}-\rho_p\right)\right)\\
&+2\sin\left(\frac{n\pi}{2}(\rho_3-\rho_2)\right)\sin\left(n\pi\left(\frac{\rho_3+\rho_2}{2}-\rho_p\right)\right)\\
&+\sin\left(\frac{n\pi}{2}(\rho_4-\rho_3)\right)\sin\left(n\pi\left(\frac{\rho_4+\rho_3}{2}-\rho_p\right)\right)\Bigg)\Bigg)
\end{aligned}
\tag{4.14}
$$

Equalizing expressions in (4.13) with the corresponding evaluation of Eq. (4.14) results in the following oscillation conditions:

$$
\begin{aligned}
\sum_{n_{odd}}^{\infty}\left(\frac{1}{n}\left(\Im\{G_{ol}(jn\omega_0)\}C_{im_1}-\Re\{G_{ol}(jn\omega_0)\}C_{re_1}\right)\right)-\frac{\pi}{4}&=0\\
\sum_{n_{odd}}^{\infty}\left(\frac{1}{n}\left(\Im\{G_{ol}(jn\omega_0)\}C_{im_2}-\Re\{G_{ol}(jn\omega_0)\}C_{re_2}\right)\right)-\frac{\pi}{2}&=0\\
\sum_{n_{odd}}^{\infty}\left(\frac{1}{n}\left(\Im\{G_{ol}(jn\omega_0)\}C_{im_3}-\Re\{G_{ol}(jn\omega_0)\}C_{re_3}\right)\right)-\frac{\pi}{4}&=0\\
\sum_{n_{odd}}^{\infty}\left(\frac{1}{n}\left(\Im\{G_{ol}(jn\omega_0)\}C_{im_4}-\Re\{G_{ol}(jn\omega_0)\}C_{re_4}\right)\right)&=0
\end{aligned},
\tag{4.15}
$$

where C_{re_k} and C_{im_k} denote the coefficients of the EC_k-th condition. These coefficients are functions solely of the values of ρ_p.

Analysis like in the $m = 1$ case cannot be undergone since $\Re\{G_{ol}(j\omega_o)\}$ and $\Im\{G_{ol}(j\omega_o)\}$ cannot be isolated from expressions in (4.15). Therefore, an analogue robustness measure to M_T cannot be obtained.

However, a more general approach can be taken. The presented conditions in (4.15) are sufficient and necessary for the induction of sustained limit cycle oscillations of $m = 2$. A robustness measure can be established as the minimal euclidean distance to the accomplishment of all the conditions. Formalizing this idea:

Definition 4.2 Consider the linear system of equations defined by (4.2) and (4.4). Expressing this system in the matrix form $Ax + B = 0$, the term L_p is defined as the left hand term of the p-th equation on that system.

Corollary 4.2 *A robustness measure against m-leveled limit cycle oscillation induction R_m is defined as:*

$$R_m = \min_{\omega,\rho_1,\cdots,\rho_{2m}} \sqrt{\sum_{p=1}^{2m} \left(L_p(\omega,\rho_1,\cdots,\rho_{2m})\right)^2} \tag{4.16}$$

Analogously to M_T, $R_m = 0$ indicates the possible induction of an m-leveled oscillation, while $R_m > 0$ indicates that such limit cycles cannot take place.

Remark 4.2 Complexity of the calculus and computation time to obtain R_m increases with m. This is due to the number of conditions to fulfill, and the number of combinations of the time fractions ρ_p to evaluate.

With the newly obtain measure R_m it can be assessed the robustness of a given system against m-leveled oscillations. The analytical demonstration of Conjecture 4.1 is not tackled due to the calculus complexity. Instead, the measure R_m has been obtained for $m = 1, 2, 3$ for the batch of processes presented in (4.17), extracted from [18], which reflect processes with the most common dynamics in actual plants.

$$\begin{aligned}
&G(s) = \frac{e^{-s}}{(Ts+1)^2},\\
&T = 0.3, 0.5, 0.7, 1, 1.3, 1.5, 2, 4, 6, 8, 10, 20, 50, 100, 200, 500\\
&G(s) = \frac{1}{(s+1)(Ts+1)^2},\\
&T = 0.05, 0.1, 0.2, 0.5, 2, 5, 10\\
&G(s) = \frac{1}{(s+1)^n},\\
&n = 3, 4, 5, 6, 7, 8\\
&G(s) = \frac{1}{(s+1)(\alpha s+1)(\alpha^2 s+1)(\alpha^3 s+1)},\\
&\alpha = 0.1, 0.2, 0.3, 0.4, 0.5, 0.6, 0.7, 0.8, 0.9\\
&G(s) = \frac{Te^{-L_1 s}}{(T_1 s+1)(Ts+1)}, \quad T_1 + L_1 = 1,\\
&T = 1, 2, 5, 10, \quad L_1 = 0.01, 0.02, 0.05, 0.1, 0.3, 0.5, 0.7, 0.9, 1\\
&G(s) = \frac{1-\alpha s}{(s+1)^3},\\
&\alpha = 0.1, 0.2, 0.3, 0.4, 0.5, 0.6, 0.7, 0.8, 0.9, 1, 1.1\\
&G(s) = \frac{1}{(s+1)((sT)^2+1.4sT+1)},\\
&T = 0.1, 0.2, 0.3, 0.4, 0.5, 0.6, 0.7, 0.8, 0.9, 1
\end{aligned} \tag{4.17}$$

For each of the processes in the batch, a controller according to [19] has been used. The obtained values for R_m have been presented in Fig. 4.3.

For the presented cases, it can be observed that the computed value for R_m increases with m. This fact does not constitute an analytical prove for Conjecture 4.1,

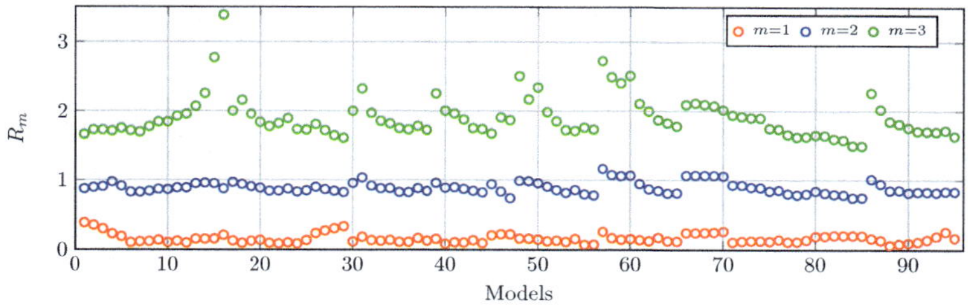

Fig. 4.3 R_1, R_2 and R_3 for the batch of models with their respective controller obtained in the previous section. (Reprinted from Journal of the Franklin Institute, Vol 357, No. 2, Miguel-Escrig, O., Romero-Pérez, J. A., & Sanchis-Llopis, R., *Tuning PID controllers with symmetric send-on-delta sampling strategy*, 832–862, Copyright (2020), with permission from Elsevier)

but empirically sustains the proposed statement, allowing the applicability of M_T for assessing robustness against limit cycle oscillations.

4.5 Evaluation of Well-Known Tuning Rules Using M_T

Among the first approaches for tuning controllers in SSOD-$C(s)$ structures like the presented in Fig. 2.1 was to employ well-known tuning rules applied to continuous control. Some of these tuning rules are Ziegler-Nichols (ZN), Cohen-Coon (CC) and AMIGO, which provide rules for both PI and PID controllers.

These methods, which have not been developed taking into account the specific loop configuration characteristics, may lead to an unexpected behavior with regard to the continuous control case.

To evaluate their performance in this kind of loop configurations, the robustness measure M_T is used in this section. The aforementioned methods have been applied to FOPTD models, known for their lack of filtering capabilities, obtaining PI and PID controllers for this kind of processes, characterized by the ratio delay (L) to time constant (τ).

The measure M_T has been obtained for the mentioned setup, and the results are presented graphically in Fig. 4.4.

In this figure, the value of M_T for each tuning rule is presented with a different color, with the PI controllers in solid line and PID in dashed line. From the presented results, it can be seen that AMIGO tuning rule offers the highest robustness for lower values of L/τ. As this ratio increases, Ziegler-Nichols results in more robust controllers. For all the tuning rules, it can be observed that it exists a critical value of the ratio L/τ from which lower values of this ratio results in oscillatory responses. Besides, this critical ratio is observed to be higher for PID than for PI controllers.

In this figure, it is also observed greater values of M_T for PI than for PID controllers of the same tuning rule. This points to the fact that PI control may be more suitable

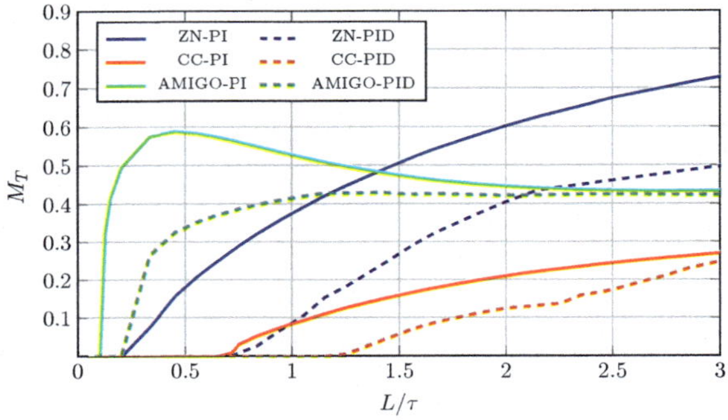

Fig. 4.4 M_T for FOPTD tuned with Ziegler-Nichols (ZN), Cohen-Coon (CC) and AMIGO tuning methods. (Adapted from IFAC-PapersOnLine, Vol 51, No. 4, Miguel-Escrig, O., Romero-Pérez, J. A., & Sanchis-Llopis, R., *New robustness measure for a kind of event-based PID*, 781–786, Copyright (2018), with permission from Elsevier)

than PID control for tuning rules not evaluated with M_T in this kind of SSOD-$C(s)$ loops.

4.6 Tsypkin's Method Approach Adapted for RQH Sampling

The developed analysis methodology can be extended to include RQH sampling. RQH sampling strategy has been presented and analyzed in Chap. 2 under the DF theory principles. As shown in this chapter, DF theory based analysis may lack accuracy in forecasting limit cycle oscillations for systems with low filtering capabilities and PID control. Therefore, in this subsection the necessary extensions to apply Tsypkin's method to RQH-$C(s)$ loops are presented.

Revisiting the previously presented analysis, Eq. (4.4) holds for describing the sampled output of a RQH sampler, which, as in the SSOD case, consists of a ladder-like signal. Nevertheless, the switching thresholds of the RQH (see equation (2.1)) differ from SSOD, resulting in different equality conditions:

$$EC_p \left| e\left(\rho_p \frac{T_o}{2}\right) = \begin{cases} \delta\left(p - \dfrac{1}{2} + \dfrac{h}{2\delta}\right) & \text{for } p = 1, 2 \ldots m \\ \delta\left(2m - p + \dfrac{1}{2} - \dfrac{h}{2\delta}\right) & \text{for } p = m+1, \ldots 2m \end{cases} \right. \tag{4.18}$$

$$IC_p \left| \frac{de\left(\rho_p \frac{T_o}{2}\right)}{dt} = \begin{cases} > 0 & \text{for } p = 1, 2 \ldots m \\ < 0 & \text{for } p = m+1, \ldots 2m \end{cases} \right. . \tag{4.19}$$

From the study presented in Sect. 4.4, it is assumed that Conjecture 4.1 holds for RQH sampling. This simplifies the analysis, leading to the following oscillation conditions:

$$EC_{p|m=1} \left| e\left(\rho_p \frac{T_o}{2}\right) = \begin{cases} \delta\left(\frac{1}{2} + \frac{h}{2\delta}\right) & \text{for } p = 1 \\ \delta\left(\frac{1}{2} - \frac{h}{2\delta}\right) & \text{for } p = 2 \end{cases} \right. \tag{4.20}$$

$$IC_{p|m=1} \left| \frac{de\left(\rho_p \frac{T_o}{2}\right)}{dt} = \begin{cases} > 0 & \text{for } p = 1 \\ < 0 & \text{for } p = 2 \end{cases} \right. . \tag{4.21}$$

As in the SSOD case, Eq. (4.4) is evaluated in $p = 1, 2$ and equated with the equality conditions in (4.20). Which ultimately leads to a new definition of the Tsypkin branches.

Definition 4.3 A Tsypkin branch $B_T(\omega, \rho)$ for RQH_ZOH samplers is defined as a trajectory in the Nyquist diagram, obtained for a given frequency ω and $\rho \in]0, 1[$, given by expression (4.22).

$$B_T(\omega, \rho) = \left(-\frac{\frac{\pi}{4} + \sum_{n=3,5..}^{\infty} \frac{1}{n} \sin(n\pi\rho)\, \Re\{G_{ol}(jn\omega)\}}{\sin(\pi\rho)}, \right.$$
$$\left. -\frac{\frac{\pi h}{8\delta} + \sum_{n=3,5..}^{\infty} \frac{1}{n} \cos^2\left(\frac{n\pi\rho}{2}\right) \Im\{G_{ol}(jn\omega)\}}{\cos^2\left(\frac{\pi\rho}{2}\right)} \right), \quad \forall \rho \in]0, 1[\tag{4.22}$$

Theorem 4.2 *A limit cycle oscillation with frequency ω_o, oscillation amplitude lesser than 2δ and commutation time fraction ρ_o will take place in RQH-$C(s)$ loops if:*

$$\exists \omega_o, \rho_o : (\Re\{G_{ol}(j\omega_o)\}, \Im\{G_{ol}(j\omega_o)\}) = B_T(\omega_o, \rho_o). \tag{4.23}$$

where $G_{ol}(j\omega)$ is the open-loop transfer function and $B_T(\omega, \rho)$ a Tsypkin branch as in Definition 4.3.

Corollary 4.3 *The Tsypkin's robustness margin M_T for RQH-$C(s)$ loops is defined as:*

$$M_T(\omega_{min}, \rho_{min}) = \min_{\omega,\rho}\{dist(G_{ol}(j\omega), B_T(\omega, \rho))\} \tag{4.24}$$

and represents the minimum euclidean distance between the open-loop transfer function and any Tsypkin branch. The analytic expression for M_T is presented in Eq. (4.25).

$$M_T(\omega_{min}, \rho_{min}) = \sqrt{\left(\frac{\frac{\pi}{4} + \sum_{n_{odd}}^{\infty} \frac{1}{n} \sin(n\pi\rho_{min}) \Re\{G_{ol}(jn\omega_{min})\}}{\sin(\pi\rho_{min})}\right)^2 + \left(\frac{\frac{\pi h}{8\delta} + \sum_{n_{odd}}^{\infty} \frac{1}{n} \cos^2\left(\frac{n\pi\rho_{min}}{2}\right) \Im\{G_{ol}(jn\omega_{min})\}}{\cos^2\left(\frac{\pi\rho_{min}}{2}\right)}\right)^2} \tag{4.25}$$

Non surprisingly, M_T results in very similar expressions for SSOD and RQH sampling. The unique a priori observable difference is in the second term of the root in (4.25). This term, contains the RQH's characteristic ratio h/δ. It can be noted that SSOD is a subcase of RQH generalization, and therefore, by considering $h = \delta$ the SSOD case can be derived from RQH expressions.

Remark 4.3 Robustness analysis for RQH-$C(s)$ loops cannot be deduced from prior SSOD-$C(s)$ analysis. That is to say, that critical pairs $(\omega_{min}, \rho_{min})$ are not kept with the change of the ratio h/δ, not being valid the direct substitution of this ratio in expression (4.25) with critical pair $(\omega_{min}, \rho_{min})$ obtained for SSOD case.

Example 4.4 Consider the process in Example 4.1 and the controller tuned according to SRB method. For illustrative purposes, consider a SSOD sampler ($h/\delta = 1$) and a RQH sampler with $h/\delta = 1/2$.

As shown in Example 4.1, this system leads to a limit cycle oscillation for the SSOD case. The analysis according to the Tsypkin method is performed for the RQH case. The graphical visualization for both cases is presented in Fig. 4.5. As it can be seen, a value of $M_T = 0.29$ is obtained for the RQH case, resulting in a stable temporal response.

The temporal response of the system under study, considering both sampling strategies, is depicted in Fig. 4.6, corroborating the analysis results.

From the representation on the Nyquist diagram presented in Fig. 4.5 it is evident that the critical oscillation frequency that defines M_T is different for both cases. This reinforces the statement in Remark 4.3, which states that robustness of RQH loops cannot be directly deduced from SSOD analysis.

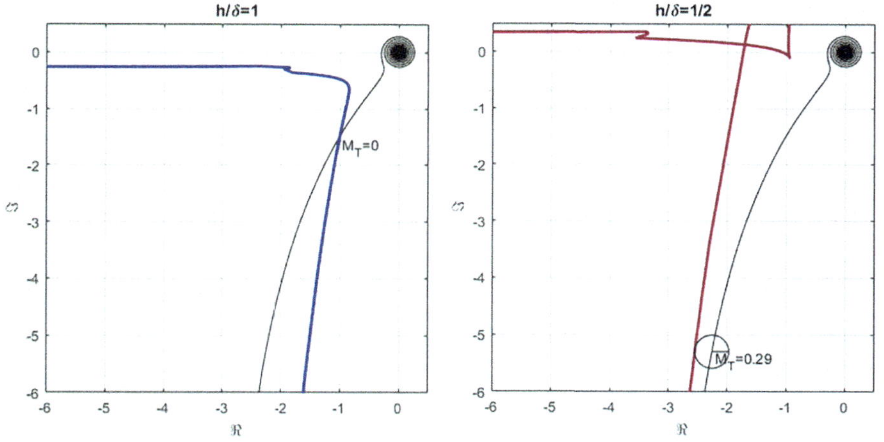

Fig. 4.5 Nyquist plot of $G_{ol}(j\omega)$ with the critical Tsypkin branch

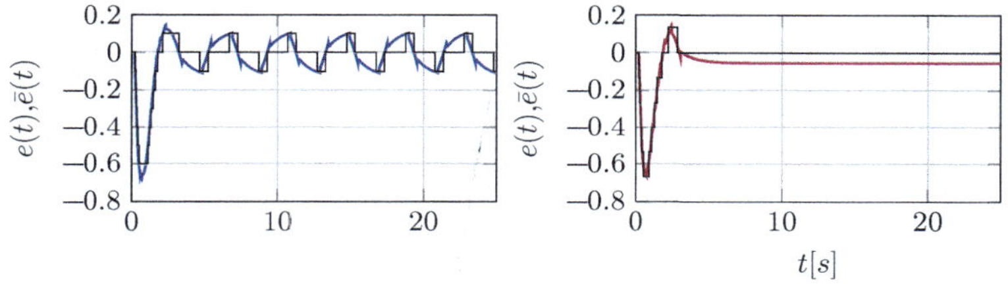

Fig. 4.6 Temporal response of the error signal $e(t)$ and sampled error signal $\bar{e}(t)$ for the studied samplers. (Same colors and disposition as in Fig. 4.5)

Quick summary:

- ☐ Robustness study of SSOD-$C(s)$ loops according to Tsypkin method has been addressed.
- ☐ A new robustness measure M_T against limit cycle oscillation induction has been defined. Empirical proves demonstrate the validity of the proposed margin through additional R_m robustness measure.
- ☐ Classical PI and PID tuning rules for continuous control have been evaluated, showing limited applicability for certain processes.
- ☐ The analysis method has been extended to RQH-$C(s)$ loops with minor changes, revealing that SSOD-$C(s)$ loops requires of controllers with stronger robustness constraints.

References

1. Laura-Marie Feeney and Martin Nilsson. Investigating the energy consumption of a wireless network interface in an ad hoc networking environment. In *Proceedings IEEE INFOCOM 2001. Conference on Computer Communications. Twentieth Annual Joint Conference of the IEEE Computer and Communications Society (Cat. No. 01CH37213)*, volume 3, pages 1548 –1557 vol.3, 2001
2. Dormido S, Sánchez J, Kofman E (2008) Muestreo, control y comunicación basados en eventos. Revista Iberoamericana de Automática e Informática Industrial RIAI 5(1):5–26
3. Joerns Ploennigs, Volodymyr Vasyutynskyy, and Klaus Kabitzsch. Comparative study of energy-efficient sampling approaches for wireless control networks. *Industrial Informatics, IEEE Transactions on*, 6(3):416 –424, aug. 2010
4. Miskowicz M (2006) Send-on-delta concept: An event-based data reporting strategy. Sensors 6(1):49–63
5. Beschi M, Dormido S, Sánchez J, Visioli A (2012) Characterization of symmetric send-on-delta PI controllers. J Process Control 22(10):1930–1945
6. Oscar Miguel-Escrig and Julio-Ariel Romero-Pérez. Regular quantisation with hysteresis: a new sampling strategy for event-based PID control systems. *IET Control Theory & Applications*, 2020
7. Chacón J, Sánchez J, Visioli A, Yebra L, Dormido S (2013) Characterization of limit cycles for self-regulating and integral processes with PI control and send-on-delta sampling. Journal of Process Control 23(6):826–838
8. Julio-Ariel Romero-Pérez, Roberto Sanchis-Llopis, and Ignacio Peñarrocha-Alós. A simple rule for tuning event-based PID controllers with symmetric send-on-delta sampling strategy. In *Proceedings of the 2014 IEEE Emerging Technology and Factory Automation (ETFA)*, pages 1–8, Sept 2014
9. Romero-Pérez J-A, Sanchis-Llopis R (2016) A new method for tuning PI controllers with symmetric send-on-delta sampling strategy. ISA Transactions 64:161–173
10. Ruiz Á, Beschi M, Visioli A, Dormido S, Jiménez JE (2017) A unified event-based control approach for FOPTD and IPTD processes based on the filtered Smith predictor. Journal of the Franklin Institute 354(2):1239–1264
11. Julio-Ariel Romero-Pérez and Roberto Sanchis-Llopis. Tuning and robustness analysis of event-based PID controllers under different event generation strategies. *International Journal of Control*, pages 1–38, 2017
12. Yakov Z (1984) Tsypkin. Cambridge University Press, Relay control systems
13. John G. Ziegler and Nathaniel B. Nichols. Optimum settings for automatic controllers. *trans. ASME*, 64(11), 1942
14. Cohen GH, Coon GA (1953) Theoretical consideration of retarded control. Transactions of the ASME 75:827–834
15. Karl J. Åström and Tore Hägglund. Revisiting the Ziegler-Nichols step response method for PID control. *Journal of Process Control*, 14(6):635–650, sep 2004
16. Manuel Beschi, Sebastián Dormido, José Sánchez, and Antonio Visioli. Tuning of symmetric send-on-delta proportional-integral controllers. *IET Control Theory & Applications*, 8:248–259(11), March 2014
17. Sanchis-Llopis R, Romero-Pérez J-A, Balaguer-Herrero P (2010) Tuning of PID controllers based on simplified single parameter optimisation. International Journal of Control 83(9):1785–1798
18. Åström Karl J, Hägglund Tore (2006) Advanced PID Control. Systems, and Automation Society, ISA-The Instrumentation
19. Miguel-Escrig O, Romero-Pérez J-A, Sanchis-Llopis R (2020) Tuning PID controllers with symmetric send-on-delta sampling strategy. Journal of the Franklin Institute 357(2):832–862

Chapter 5
Sampled SSOD Analysis with DF Method

5.1 Introduction

Event-Based Control (EBC) serves as a robust alternative to traditional time-driven control in distributed control systems. This is achieved by mitigating data dropouts, minimizing delays, and reducing packet losses in communication networks. EBC accomplish this through their data transmission policy, sending updates only when substantial changes in the system state occur, in contrast to the periodic transmissions of classical time-driven controllers.

Hence, the data transmission policy holds significant importance in EBC as it is responsible for issuing events for the execution of the controller's algorithm. Among the various event generation techniques, those relying on signal quantification have gained prominence due to their simplicity of implementation. An example is the well-known Send-On-Delta (SOD) sampling strategy, which transmits data whenever the signal undergoes a change exceeding a specified value, denoted as δ, from the last sampled value. Numerous studies have utilized this sampling technique, demonstrating its effectiveness in enhancing control performance and reducing communication load [1, 2].

Numerous adaptations of the SOD sampling strategy have been introduced, with many maintaining fixed thresholds independent of the previous value sampled. Regular Quantization, whose robustness was explored in [3], is one such variation. Another variant was proposed in [4] (see Chap. 2), introducing customizable hysteresis to mitigate event bursts caused by noise. However, one of the most recognized modifications is Symmetric-Send-On-Delta (SSOD), as highlighted in [5]. In SSOD, fixed thresholds with a value δ are established, accompanied by an identical δ-valued hysteresis, resulting in a symmetric input-output relationship.

In the analysis and design of EBC systems, a crucial aspect involves identifying and mitigating limit cycle oscillations in the closed-loop response. Steady-state oscillations characterization under a SSOD sampling strategy has been explored in [6]. Previous works, such as [7, 8], have employed the Describing Function (DF)

O. Miguel-Escrig, *Event-Based PID Controllers with Fixed Threshold Sampling Strategies*, Springer Theses, https://doi.org/10.1007/978-3-031-62505-3_5

technique for this analysis. In addition to its use in analysis, the DF technique has been applied to obtain tuning methods for PI controllers within SSOD-PI control architectures. The DF technique extends classical control theory concepts, such as gain and phase margins, to the analysis and design of EBC systems. However, alternative analysis techniques, such as the Tsypkin method, have also been utilized, as demonstrated in [9] (see Chaps. 4 and 7), specifically for assessing the robustness of SSOD-PID structures.

In all theoretical analyses of SSOD-based control systems, a continuous approach has traditionally been taken towards controller implementation. That is, the controller is assumed to be continuous, and all outcomes are derived under this assumption. In contrast, in networked control systems, controllers are implemented in microprocessor-based devices using a discrete approximation. This chapter aims to investigate the impact of sampling time on robustness against limit cycle oscillations when employing a discrete implementation of the PI controller in conjunction with the SSOD sampling strategy. This scenario reflects practical applications of SSOD-PI controllers, as illustrated in [10–14]. Remarkably, none of these previous papers have delved into the influence of this crucial parameter.

Control practitioners often employ tuning methods designed for continuous PID controllers in various industrial settings. However, it is crucial to recognize that these controllers are executed in microprocessor-based systems as discrete versions. In many instances, the digital implementation does not adversely impact the loop performance because the control algorithm's sampling frequency is high enough to treat the controller as continuous. For common process variables like flow, level, pressure, or temperature, the sampling time required is in the order of seconds. Meanwhile, commercial digital controllers are set with sampling intervals in the order of milliseconds.

Despite the aforementioned consideration, it is widely acknowledged that the sampling time plays a crucial role in discrete-time control systems. Elevated values of sampling time have the potential to deteriorate control performance or even induce instability. Conversely, the minimum sampling interval is constrained by the hardware and software limitations of the microprocessor-based system executing the control algorithms. Due to the significance of sampling time, various criteria have been developed to guide the selection of this parameter, considering both the frequency and time response of the control loop [15]. A commonly recommended practice for PI controllers suggests setting the sample time within the range of 0.1–0.3 times the integral time to achieve optimal performance and acceptable robustness in control systems [16].

This chapter delves into the robustness of SSOD based control systems when a discrete PI controller is employed. This consideration aligns with the prevalent reality of computer-based controller implementations, which typically function as periodic algorithms. The analysis is grounded in the Sampled Describing Function [17], and encompasses the effects of SSOD quantification and the execution period of the controller. Key characteristics of this DF have been analyzed, unveiling the impact of its parameters on the behavior of such control systems. The chapter introduces guidelines for assessing the presence of limit cycles using a novel robustness measure.

Due to the absence of dedicated tuning rules for the scheme under examination, the robustness of controllers using various classical and widely-used tuning rules for continuous PI controllers has been evaluated. These include the Ziegler-Nichols [18], AMIGO [19], One-Third [20], and SIMC [21] tuning rules. These methods have been applied to a set of models encompassing the most common dynamics in industrial processes. Their robustness within the studied loop structure has been assessed using the proposed measure.

5.2 Problem Statement

Consider the networked control system presented previously in Chap. 4 based on Fig. 2.1 and its adaptation for analysis presented in Fig. 4.1. Description of the main elements can be found in the aforementioned chapter.

Studies using this loop structure, see for instance [5, 7–9, 22], have approached the stability and tuning problem in several manners, but none of them have considered a discrete implementation of the controller.

This simplification may lead to unexpected behaviors when implementing the controllers in computer-like devices like PLCs or other electronic cards. Therefore, the controller is now considered discrete, characterized by:

$$C(z) = K_p + \frac{K_p}{T_i}\frac{T_s}{2}\frac{z+1}{z-1}, \tag{5.1}$$

obtained by applying the bilinear transform. Besides, a Zero-Order Hold following the controller is added.

As a consequence of this change, the study of the induction of limit cycle oscillations is also modified. Now, the error signal e follows the quantification by the SSOD and, then, it is sampled, resulting in the sampled error signal $\bar{e}^*$, which is the input to the controller.

To study the effects on the robustness against these limit cycle oscillations, an analysis methodology is developed using the Sampled Describing Function.

5.3 Sampled Describing Function Approach

The effect of sampling elements on the robustness of the loop under study can be evaluated through several methods. However, exact methods like [23] result complicated for practical uses. Thus, other methods are preferred, like those based on the Describing Function theory, which provide an approximated estimation of the robustness while being easy to use. In this sense, two variants exist to analyze sampled systems, the z-transform Describing Function [24] and the Sampled Describing Function [17]. This last method is used for its simplicity and documentation [25].

DF theory defines the following condition to avoid limit cycle oscillations:

$$G_{ol}(j\omega) \neq -\frac{1}{\mathcal{N}}; \quad \forall\omega, \tag{5.2}$$

where $G_{ol}(j\omega)$ is the open-loop transfer function of all the linear elements in the loop and $\mathcal{N}$ is the describing function of the non-linear part.

The expression of $\mathcal{N}$ for the sampled SSOD as described in this chapter is given by the following expression (see Appendix B.4):

$$\mathcal{N}(\delta/A, r, \tau, T_s) = \frac{2\delta}{\pi T_s A} \sum_{k=-\infty}^{\infty} \frac{e^{-jk2\pi\tau/T_s}}{rk-1} \left[\sum_{i=1}^{m-1} \left\{ \left(-\sqrt{1-\left(\frac{i\delta}{A}\right)^2} + j\frac{i\delta}{A} \right)^{rk-1} - \left(\sqrt{1-\left(\frac{i\delta}{A}\right)^2} + j\frac{i\delta}{A} \right)^{rk-1} \right\} - \left(\sqrt{1-\left(\frac{m\delta}{A}\right)^2} + j\frac{m\delta}{A} \right)^{rk-1} + (-1)^{rk-1} \right] \tag{5.3}$$

where $m = \left\lfloor \frac{A}{\delta} \right\rfloor$ is the number of thresholds crossed, τ is the elapsed time between the origin and the first sample taken, and r is the ratio oscillation period T_o to sampling period T_s.

As in the case of the continuous DF of the SSOD sampling [8], the expression given in (5.3) depends on the ratio δ/A and not on each value independently. Besides, this expressions presents new parameters related to the discrete case.

One of those parameters that is worth mentioning is the ratio r, which according to [25], is an integer. If this is not fulfilled the resulting oscillation may contain harmonics lower than the observed fundamental frequency, which would invalidate the applicability of the DF. Besides, the cases where r is odd result in an asymmetry which is relevant in the cases where r is low, otherwise its effect is unnoticeable.

In Fig. 5.1 the shape of $-1/\mathcal{N}$ for different values of r is depicted. Several branches compose the locus defined by $-1/\mathcal{N}$, each of them for a different value of m, which have been represented with a different color. As m increases, the traces of the DF tend towards the point $(-T_s, 0)$. In Fig. 5.1d is found the most representative example of the stated above, reassembling to the continuous DF for the SSOD (see [8]). This is observed since the ratio r is high enough to neglect the sampling effect. As r decreases the branches spread and overlap.

The concept of oscillation mode is introduced for a relay non-linearity in [25]. In that context, a n, n mode is defined as a cycle comprising n positive drive pulses followed by n negative drive pulses. This concept can be adapted to the oscillations generated by the non-linear structure examined in this chapter, with a slight modification to accommodate the specific characteristics of the non-linearity.

Taking as an example a single-leveled oscillation ($m = 1$), consider $r = T_o/T_s = 6$, that is six samples in an oscillation period. In each semiperiod, the two possible

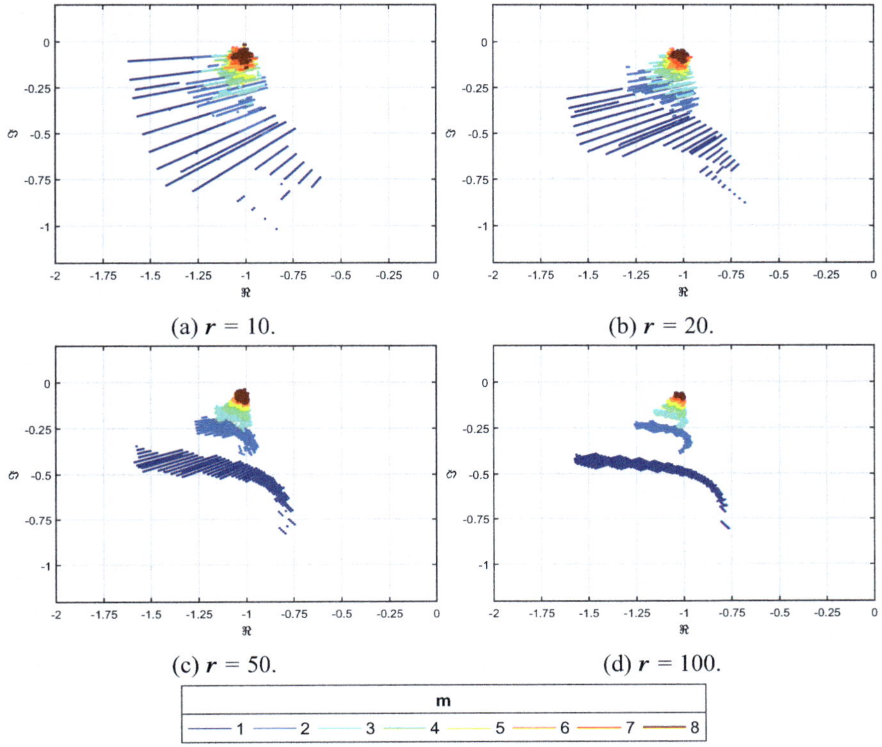

(a) $r = 10$.

(b) $r = 20$.

(c) $r = 50$.

(d) $r = 100$.

Fig. 5.1 Sampled DF for different values of r, all of them considering $T_s = 1$. (Reprinted from International Journal of Control, Vol. 97, No. 1, Miguel-Escrig, O., & Romero-Pérez, J. A., *Event-based discrete PI controllers robustness analysis through sampled describing function technique*, 86–100, Copyright (2024), with permission from Taylor & Francis Ltd, http://www.tandfonline.com.)

combinations of samples are $\bar{e}^*(kT_s) = 0, 0, \delta$ or $\bar{e}^*(kT_s) = 0, \delta, \delta$ for $k = 1, 2, 3$, which constitute the oscillation modes. It is worth noting that on the negative semiperiod the samples will be the same but with opposite sign.

Considering a given ratio r, the number of modes in an induced single-leveled sampled oscillation can be computed as:

$$n^\circ modes = \left\lceil \frac{r}{4} \right\rceil - \left\lfloor \frac{r}{12} \right\rfloor + 1. \tag{5.4}$$

Proof Sinusoidal oscillations in e inducing single-leveled sampled oscillations have an amplitude $A \in [\delta, 2\delta[$. For these extreme cases the commutation is produced at $t_1 = T_o/4$ and $t_2 = T_o/12$ respectively. The number of oscillation modes is the different combinations of samples that fit, which is obtained by dividing the difference with the times above by T_s, thus, appearing the ratio r. Rounding functions are included for the extreme cases. □

The effect of r in the oscillation modes is also reflected graphically in the DF traces. For instance, in Fig. 5.1c, several rhomboid regions appear, which are more visible for the $m = 1$ trace. Each of these regions corresponds to a different oscillatory mode, and each straight line across the rhomboid a different initial τ.

Since SSOD sampling encompasses more than one oscillation levels, more oscillatory modes can be found when $m > 1$. The complexity in those cases escalates due to the increasing number of combinations, resulting in overlapping regions for $m > 1$ in the Nyquist diagram. However, this is not a troublesome issue since in this chapter the focus is on $m = 1$ oscillations.

Finally, it is important to observe that the variation in the DF traces with respect to the parameter r can be elucidated by considering the concept of modes. A lower ratio of r results in a limited number of modes and a greater range of parameter combinations that can achieve them. This leads to a pronounced dispersion in the DF traces. Conversely, higher ratios of r increase the number of modes (represented as rhomboid regions), but these regions are smaller due to the constrained parameter variations required to sustain them.

5.4 Stability Analysis

A key aspect in examining the networked control systems discussed in this chapter is the prevention of limit cycles introduced by the SSOD. These oscillations are avoided when the condition in Eq. (5.2) is satisfied. This condition establishes a relationship between the non-linear component, characterized by the DF $\mathcal{N}$, and the linear part of the system.

Regarding to the linear part of the system, the open-loop transfer function $G_{ol}(s)$ is composed of the process and network latency $G(s)$, the controller $C(z)$ and the ZOH, which has the following transfer function:

$$ZOH(s) = \frac{1 - e^{-sT_s}}{s}.$$

For the controller, the transformation $z = e^{sT_s}$ is applied to the expression in (5.1), resulting in the starred transform of the controller $C^*(s)$. The open-loop transfer function results in the expression:

$$G_{ol}(s) = C^*(s)ZOH(s)G(s). \tag{5.5}$$

The stability condition (5.2) has been rewritten as follows to facilitate the analysis:

$$\frac{G_{ol}(s)}{T_s} \neq \frac{-1}{\mathcal{N}'}; \quad \forall\omega, \tag{5.6}$$

where:

$$\mathcal{N}'(\delta/A, r, \theta_s) = \frac{2\delta}{\pi A} \sum_{k=-\infty}^{\infty} \frac{e^{-jk\theta_s}}{rk-1} \left[\sum_{i=1}^{m-1} \left\{ \left(-\sqrt{1-\left(\frac{i\delta}{A}\right)^2} + j\frac{i\delta}{A} \right)^{rk-1} - \left(\sqrt{1-\left(\frac{i\delta}{A}\right)^2} + j\frac{i\delta}{A} \right)^{rk-1} \right\} - \left(\sqrt{1-\left(\frac{m\delta}{A}\right)^2} + j\frac{m\delta}{A} \right)^{rk-1} + (-1)^{rk-1} \right], \tag{5.7}$$

which is the same expression than in (5.3) but without the sampling period, which has been moved to the left term of equation (5.6). In addition, the term $2\pi\tau/T_s$ has been replaced by θ_s, since $\tau \in [0, T_s[$, then $\theta_s \in [0, 2\pi[$.

These adjustments render $\mathcal{N}'$ dimensionless and eliminate its scaling by T_s. Consequently, akin to the continuous DF for SSOD, $\mathcal{N}'$ converges toward $(-1, 0)$ as m increases, independently of the sampling period.

Furthermore, for the sake of uniform notation in the stability condition (5.6), the frequency ω is expressed in terms of r and the sampling frequency ω_s:

$$\frac{G_{ol}\left(j\frac{\omega_s}{r}\right)}{T_s} \neq \frac{-1}{\mathcal{N}'(\delta/A, r, \theta_s)}; \quad \forall r \in \mathbb{Z},\ r_{min} < r < r_{max} \tag{5.8}$$

In this expression the oscillation frequency appears in terms of the sampling ratio ($r = T_o/T_s = \omega_s/\omega_o$) since each evaluated frequency is candidate to fulfill the condition. The resulting expression (5.8) depends on the evaluated frequency indirectly on both sides of the expression, thus, the range for ω must be obtained.

Firstly, as in the continuous DF for SSOD, the traces of the inverse negative of the Sampled DF are placed in the third quadrant of the Nyquist plot (see Fig. 5.1). Therefore, only frequencies of $G_{ol}(j\omega)$ must be evaluated. Guidelines for the selection of the frequency range to evaluate are provided in the following lines.

Considering a given sampling frequency ω_s, the frequency band to evaluate is bounded through the single parameter r. The minimum value for r will correspond to the maximum frequency in the third quadrant. As the crossover phase frequency ω_{cp} corresponds to the limit case, the minimum value of r is determined as:

$$r_{min} = \left\lceil \frac{\omega_s}{\omega_{cp}} \right\rceil. \tag{5.9}$$

Conversely, the maximum value for r can be retrieved studying the DF traces. It is important to note that, with a fixed sampling frequency, increasing values of r correspond to lower frequencies ω_o. Thus, increasing r results in traces of $\mathcal{N}'$ reassembling the continuous DF traces. Therefore, the robustness study can be limited to those frequencies that lead to a magnitude of G_{ol}/T_s in a range corresponding to the magnitudes in which $-1/\mathcal{N}'$ would be placed. The highest magnitude for the traces has been measured to be 1.62 units. Thus, points of $G_{ol}(s)/T_s$ with magnitude higher to this value are not evaluated.

The maximum value for r can be obtained analytically as:

$$r_{max} = \left\lceil \frac{\omega_s}{\omega'_{cg}} \right\rceil, \tag{5.10}$$

being ω'_{cg} the frequency of $G_{ol}(j\omega)/T_s$ leading to the maximum modulus that the DF traces reach:

$$\left\| \frac{G_{ol}(j\omega'_{cg})}{T_s} \right\| = 1.62. \tag{5.11}$$

An example illustrating the analysis procedure for condition in (5.8) is presented.

Example 5.1 Consider a process defined by the following transfer function:

$$G(s) = \frac{1}{(s+1)^3}.$$

A controller tuned according to [8] is tuned. This method offers robust controllers in control schemes like the presented in Fig. 2.1, which corresponds to a continuous SSOD-$C(s)$ loop. A controller fulfilling $\Phi_{m,SSOD} = 10°$ and $\gamma_{cp} \geq 6$ dB is obtained with $K_p = 1.35$ and $T_i = 2.38$, not inducing limit cycle oscillations for the continuous implementation of the controller.

Aiming to analyze the robustness for a discrete implementation of the controller, a sampling period of $T_s = 0.5$ s is considered.

The minimum and maximum values for r are now determined. According to expression (5.9) the crossover phase frequency is measured ($\omega_{cp} = 1.11$ rad/s) resulting in $r_{min} = \lceil \omega_s/\omega_{cp} \rceil = \lceil 12.57/1.11 \rceil = 12$. Conversely, the crossover gain frequency of $G_{ol}(s)/1.62/T_s$ is obtained ($\omega'_{cg} = 0.38$ rad/s.), and according to expression (5.10), $r_{max} = \lceil \omega_s/\omega'_{cp} \rceil = \lceil 12.57/0.38 \rceil = 33$.

In Fig. 5.2a the Nyquist diagram of $G_{ol}(s)/T_s$ and the traces of the Sampled DF for $m = 1$ are shown. For several values of r, the corresponding points of the Sampled DF $-1/\mathcal{N}'(r)$ are presented in different colors.

According to (5.8), limit cycle oscillations are induced if any frequency response point and DF trace calculated for the same r intercept, in the figure, points with the same color. Since the DF traces overlap and do not offer a good visibility, the traces of $-1/\mathcal{N}'(r)$ for each color presented in Fig. 5.2a have been replaced by their respective convex hull in Fig. 5.2b. With this change in the representation, a limit cycle oscillation is induced if $G_{ol}(j\omega_s/r)/T_s$ lies within the convex hull generated by $-1/\mathcal{N}'(r)$ traces.

In this example, $G_{ol}(j\omega_s/20)$ lies within the convex hull generated by $-1/\mathcal{N}'(20)$ traces, thus, limit cycle oscillations with $r = 20$ may be inducted. Simulation of this system reveals that limit cycle oscillations are inducted for the discrete implementation of the controller. Frequency of these oscillations has been measured, resulting in $\omega_s/20$, confirming the results observed in Fig. 5.2b, which shows that exactly that frequency is placed within its convex hull.

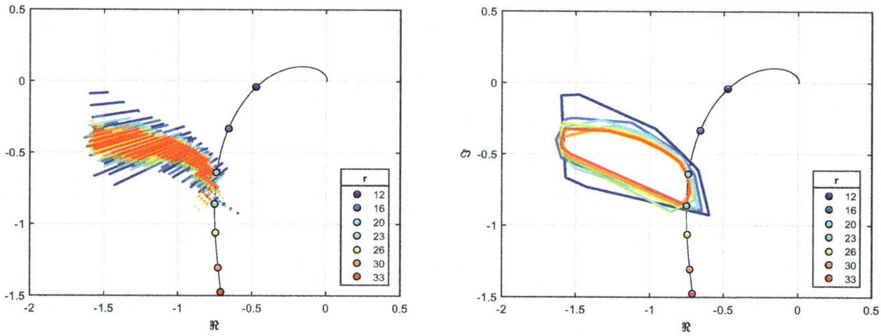

(a) Original $-1/\mathcal{N}'$ representation. (b) $-1/\mathcal{N}'$ replaced by its respective convex hull.

Fig. 5.2 Sampled DF for $m = 1$ with $T_s = 0.5$ considering different values of r. (Reprinted from International Journal of Control, Vol. 97, No. 1, Miguel-Escrig, O., & Romero-Pérez, J. A., *Event-based discrete PI controllers robustness analysis through sampled describing function technique*, 86–100, Copyright (2024), with permission from Taylor & Francis Ltd, http://www.tandfonline.com.)

5.4.1 Robustness Measure

As previously shown in Fig. 5.1, branches of $-1/\mathcal{N}'(r)$ shrink and are placed closer to the real axis as m increases. That is the reason why in Example 5.1 only $m = 1$ traces were considered.

Generally speaking, the shape of $G_{ol}(s)/T_s$ with a proper tuning is such that its distance with regard with $-1/\mathcal{N}'(r)$ increases with m. Thus, avoiding intersection with $m = 1$ branches also guarantees no intersection with $m > 1$ branches. Studies like [8, 9] point in this direction for the continuous implementation of the controller in SSOD-$C(s)$ loops. Therefore, only $m = 1$ branches will be considered for robustness definition.

Assume that Conjecture 4.1 holds for the discrete implementation of PI controllers with reasonable tuning. Consider the stability criterion defined in expression (5.8), a robustness measure for SSOD-$C(z)$ loops, named D_{ch}, is defined as the minimum distance between each point of the frequency response and the normalized Sampled DF in the Nyquist diagram:

$$D_{ch} = \min_{r_{min} \leq r \leq r_{max}} \left\{ \text{dist} \left(\frac{G_{ol}\left(\frac{j\omega_s}{r}\right)}{T_s}, CH\left(-\frac{1}{\mathcal{N}'(r)}\right) \right) \right\} \tag{5.12}$$

dist$(\cdot, \cdot)$ is the Euclidean distance between two pints and $CH(\cdot)$ the convex hull defined by the set in the argument.

The proposed margin is illustrated in the following example.

Example 5.2 Consider a process modeled by the following transfer function:

$$G(s) = \frac{1}{(s + 1)^5}$$

Parameters for a PI controller have been obtained through AMIGO tuning rule [19] resulting in $K_p = 0.2564$ and $T_i = 2.891$. A sampling period of $T_s = 1.9$ has been chosen for the discretization of the controller. It corresponds to a tenth of the continuous step response's rise time. A trapezoidal implementation is chosen for the controller.

The robustness analysis has been performed as described in the previous section. The resulting convex hulls for each of the evaluated values of r and the corresponding frequency response points are presented in Fig. 5.3a. For each value of r, a circle is presented, which is the minimum distance to each convex hull. Among these circles, the one that defines D_{ch}, is obtained for $r = 31$. To facilitate the visualization, the convex hull defining D_{ch} is highlighted in Fig. 5.3b, where the remaining distances to convex hulls are presented in gray.

Since a magnitude $D_{ch} = 0.45$ has been obtained for the studied example, the temporal response will avoid limit cycles. Simulations using step-like changes in reference and disturbance inputs corroborated the analysis results.

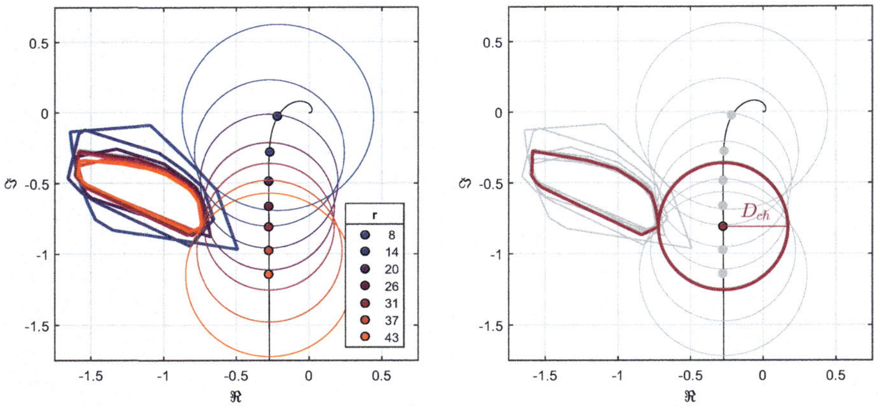

(a) All the distances to convex hull. (b) D_{ch} highlighted.

Fig. 5.3 Representation of the distances from the evaluated points to their respective convex hull and detail of D_{ch} for the studied system. (Reprinted from International Journal of Control, Vol. 97, No. 1, Miguel-Escrig, O., & Romero-Pérez, J. A., *Event-based discrete PI controllers robustness analysis through sampled describing function technique*, 86–100, Copyright (2024), with permission from Taylor & Francis Ltd, http://www.tandfonline.com.)

5.5 Robustness of Continuous Tuning Rules

In this section, well-known tuning rules for continuous control, namely, Ziegler-Nichols [18], AMIGO [19], One-Third rule [20] and SIMC [21], are evaluated through the proposed robustness margin. PI controllers according to these methods have been tuned for each model presented in the batch in (5.13), which gather a wide range of actual systems. To obtain the parameters, the dynamic responses of these models are approximated by a FOPTD model. Controllers are implemented in the trapezoidal form as shown in expression (5.1).

Sampling periods for each controller and process have been obtained as a fraction of the continuous rise time T_r. Firstly, the measure D_{ch} for each system and controller in the batch has been obtained considering $T_s = T_r/10$.

Results with this sampling period indicate that controllers tuned with Ziegler-Nichols method for the presented batch tend to present an oscillatory response. The rest of the analyzed tuning rules present distinct degrees of robustness regarding to D_{ch} but all of them consistently avoid oscillations.

$$
\begin{aligned}
G(s) &= \frac{e^{-s}}{(Ts+1)^2},\\
T &= 0.01, 0.02, 0.05, 0.1, 0.2, 0.3, 0.5, 0.7, 1,\\
&\quad 1.3, 1.5, 2, 4, 6, 8, 10, 20, 50, 100, 200, 500\\[1em]
G(s) &= \frac{1}{(s+1)(Ts+1)^2},\\
T &= 0.05, 0.1, 0.2, 0.5, 2, 5, 10\\[1em]
G(s) &= \frac{1}{(s+1)^n},\\
n &= 3, 4, 5, 6, 7, 8\\[1em]
G(s) &= \frac{1}{(s+1)(\alpha s+1)(\alpha^2 s+1)(\alpha^3 s+1)},\\
\alpha &= 0.1, 0.2, 0.3, 0.4, 0.5, 0.6, 0.7, 0.8, 0.9\\[1em]
G(s) &= \frac{Te^{-L_1 s}}{(T_1 s+1)(Ts+1)},\\
T_1 + L_1 &= 1, \quad T = 1, 2, 5, 10\\
L_1 &= 0.01, 0.02, 0.05, 0.1, 0.3, 0.5, 0.7, 0.9, 1\\[1em]
G(s) &= \frac{1-\alpha s}{(s+1)^3},\\
\alpha &= 0.1, 0.2, 0.3, 0.4, 0.5, 0.6, 0.7, 0.8, 0.9, 1, 1.1\\[1em]
G(s) &= \frac{1}{(s+1)((sT)^2+1.4sT+1)},\\
T &= 0.1, 0.2, 0.3, 0.4, 0.5, 0.6, 0.7, 0.8, 0.9, 1
\end{aligned}
\tag{5.13}
$$

From these methods, AMIGO and SIMC offer uniform levels of robustness through the batch. One-Third tuning rule presents a drop of robustness for some systems because of the fact that the sampling frequency is in the order of magnitude of the crossover phase frequency, resulting in low values for r_{min} (see expression 5.9). These low values of r, as shown in Fig. 5.1, produce a dispersion of $-1/N'(r)$ traces, enlarging the corresponding convex hull, which becomes the critical case defining D_{ch}. Nevertheless, One-Third tuning method also avoids consistently limit cycle induction.

A second study has been conducted considering this time $T_s = T_r/20$. The obtained results indicate a similar behavior as for the precedent study with some improvements in robustness. Ziegler-Nichols method offer the lowest robustness among all the methods, and some systems would present limit cycle oscillations since $D_{ch} = 0$ is measured. AMIGO and SIMC offer uniform and consistent robustness through all the batch. One-Third rule continues exhibiting bumps in robustness but are less frequent.

Comparing both studies, the results for the ad-hoc measure D_{ch} match with classical discrete control: increasing the sampling frequency results in a rise of robustness. The average increase through all the systems is of $\Delta D_{ch} = 0.073$ units. Besides, by increasing the sampling frequency, the minimum value for r has also increased, which leads to more focused and restrained convex hulls (see Fig. 5.1).

Remark 5.1 Generally speaking, there is no established value for D_{ch} that always guarantees a favorable behavior of the closed-loop response. Nevertheless, from the previous studies, values of D_{ch} within the range of 0.2-0.6 have been derived. Given that certain of the previous PI controllers yield satisfactory temporal responses, values of D_{ch} within this range could be deemed reasonable for achieving commendable closed-loop performances.

5.5.1 Effect of Sampling Period on the Performance

Discrete controllers tend to exhibit lower performance compared to continuous control systems. This is often attributed to the reduced information content in sampled signals compared to continuous signals. To investigate the impact of sampling period on the performance of the discrete event-based control loop under consideration, the index IAE^* has been employed, defined as:

$$IAE^* = \frac{\overline{IAE}}{IAE_c}$$

$\overline{IAE}$ and IAE_c denote the IAE indexes against step disturbance for the discrete EBC and continuous controller respectively.

Different sampling rates with regard to the rise time have been considered for the study, which encompasses the processes and controllers from the previous section: $T_s = T_r/20$, $T_s = T_r/10$ and $T_s = T_r/3$.

Results show that there is not a major difference in performance between controllers with $T_s = T_r/10$ and $T_s = T_r/20$. This contrasts with the notable loss in robustness reported in previous section. Besides, by considering $T_s = T_r/3$ a decrease in performance has been measured for almost every case when comparing with precedent sampling rates. This is influenced by the limited robustness against limit cycles, measured through D_{ch}, for the studied systems, which for most cases presented an oscillatory behavior ($D_{ch} = 0$). These facts discourage the usage of this sampling rate.

Therefore, it is proven the paramount importance of the sampling rate in performance and robustness, which are both enhanced when increasing the sampling rate.

5.5.2 Influence of Model Uncertainties on D_{ch}

Concerning uncertainties in plant parameters, it is logical that a higher value of D_{ch} allows for greater admissible modeling errors or variations in parameters before limit cycle oscillations are inducted. Uncertainties in the model can impact both the magnitude and phase of G_{ol}. When uncertainties affect the magnitude of G_{ol}, such as those related to process gain, they lead to a radial movement of $G_{ol}(s)/T_s$. This displacement expands or contracts as the gain increases or decreases, respectively. Therefore, since the convex hulls utilized for calculating D_{ch} remain the same, increasing the process gain results in a reduction of D_{ch} because the critical points move closer to their respective convex hull.

Uncertainties affecting the phase of the process modify the crossover phase frequency, thereby altering the value of r_{min}. This, in addition, induces a rotation of $G_{ol}(s)/T_s$ on the Nyquist diagram, causing the critical points to either approach or move away from their respective convex hulls. A phase increment can be interpreted as an increasing time delay, which introduces ωL radians to the phase of $G_{ol}(s)/T_s$, where L represents the delay. In such a scenario, the robustness measure D_{ch} would decrease as lower values of r are evaluated, resulting in increased dispersion of the Sampled DF traces. Besides, the rotation on $G_{ol}(s)/T_s$ would bring the critical points closer to their convex hulls.

Additionally, variations in the open-loop transfer function can be induced by fluctuations in model parameters like zeros and poles. Despite these general considerations, a detailed analysis specific to each parameter's admissible uncertainties in a particular model must be conducted.

5.5.3 Multi-leveled Oscillations

The conducted study based on the measure D_{ch} relies on the fulfillment of Conjecture 4.1. For the continuous case, this conjecture has been studied in [8, 9], but it has not been addressed yet for the sampled case.

The proposed robustness measure D_{ch} is flexible enough to be adapted to the multi-leveled oscillation case. To that end, the corresponding traces of $-1/\mathcal{N}'$ of the level under study in Eq. (5.12) must be studied.

In this section, the robustness against two leveled limit cycles have been studied, that is, considering $m = 2$ in Eq. (5.7). As explained in previous sections, the range of r has been correspondingly obtained, then the matching traces resulting to the convex hull and, ultimately, the adapted robustness measure $D_{ch}(m = 2)$.

This adapted measure has been obtained for the process in the batch in (5.13) considering the controllers tuned according to Ziegler-Nichols, AMIGO, One-Third and SIMC tuning rules.

Comparing the adapted measure with the robustness measure for single-leveled oscillations $D_{ch}(m = 1)$, an increase of robustness has been observed in all the cases. Exceptions for this are found in the oscillatory systems for $m = 1$, which also oscillate for $m = 2$. Therefore, this study constitutes an additional probe of the veracity of Conjecture 4.1, which now is also extended to the sampled case.

Quick summary:

- ☐ A robustness analysis for discrete event-based systems, using SSOD as quantifier, is presented.
- ☐ The Sampled Describing Function is employed for the analysis. For the extreme case, results agree with the continuous DF.
- ☐ Several modifications are applied to the classical Sampled DF approach, conducting to the robustness measure against limit cycles for these kind of systems D_{ch}.
- ☐ Well-known tuning rules for continuous control are evaluated, offering diverse levels of robustness depending on the tuning method and process.
- ☐ The study indicates that controller robustness increases with higher sampling frequencies, aligning with sampled control theory principles.
- ☐ The effect of the sampling rate on the performance and the influence of model uncertainties on the robustness has been evaluated.

References

1. Dormido S, Sánchez J, Kofman E (2008) Muestreo, control y comunicacián basados en eventos. Revista Iberoamericana de Automática e Informática Industrial RIAI 5(1):5–26
2. Ploennigs J, Vasyutynskyy V, Kabitzsch K (2010) Comparative study of energy-efficient sampling approaches for wireless control networks. IEEE Trans Ind Inf 6(3):416 –424
3. Romero-Pérez J-A, Sanchis-Llopis R (2017) Tuning and robustness analysis of event-based PID controllers under different event generation strategies. Int J Control 1–38
4. Miguel-Escrig O, Romero-Pérez J-A (2020) Regular quantisation with hysteresis: a new sampling strategy for event-based PID control systems. IET Control Theory Appl
5. Beschi M, Dormido S, Sánchez J, Visioli A (2012) Characterization of symmetric send-on-delta PI controllers. J Process Control 22(10):1930–1945
6. Chacón J, Sánchez J, Visioli A, Yebra L, Dormido S (2013) Characterization of limit cycles for self-regulating and integral processes with PI control and send-on-delta sampling. J Process Control 23(6):826–838
7. Romero-Pérez J-A, Sanchis-Llopis R, Peñarrocha-Alós I (2014) A simple rule for tuning event-based PID controllers with symmetric send-on-delta sampling strategy. In: Proceedings of the 2014 IEEE emerging technology and factory automation (ETFA), Sept 2014, pp 1–8
8. Romero-Pérez J-A, Sanchis-Llopis R (2016) A new method for tuning PI controllers with symmetric send-on-delta sampling strategy. ISA Trans 64:161–173
9. Miguel-Escrig O, Romero-Pérez J-A, Sanchis-Llopis R (2020) Tuning PID controllers with symmetric send-on-delta sampling strategy. J Frank Inst 357(2):832–862
10. Beschi M, Pawlowski A, Guzmán JL, Berenguel M, Visioli A (2014) Symmetric send-on-delta pi control of a greenhouse system. IFAC Proc 47(3):4411–4416. 19th IFAC World Congress
11. Romero JA, Sanchis R, Arrebola E (2015) Experimental study of event based pid controllers with different sampling strategies. Application to brushless dc motor networked control system. In: 2015 XXV international conference on information, communication and automation technologies (ICAT), pp 1–6
12. Beschi M, Dormido S, Sanchez J, Visioli A, Yebra LJ (2014) Event-based pi plus feedforward control strategies for a distributed solar collector field. IEEE Trans Control Syst Technol 22(4):1615–1622
13. Rodríguez-Miranda E, Beschi M, Guzmán JL, Berenguel M, Visioli A (2019) Application of a symmetric-send-on-delta event-based controller for a microalgal raceway reactor. In: 2019 18th European control conference (ECC), pp 3132–3137
14. Rodríguez-Miranda E, Beschi M, Guzmán JL, Berenguel M, Visioli A (2019) Daytime/nighttime event-based PI control for the PH of a microalgae raceway reactor. Processes 7(5)
15. Isermann R (1989) Digital control systems, volume 1 fundamentals, deterministic control, chapter choice of sample time for parameter-optimized control algorithms, 2nd ed, pp 151–154. Springer, Berlin, Heidelberg GmbH
16. Astrom KJ, Wittenmark B (1997) Computer-Controlled systems. Theory and design. Prentice Hall, 3rd ed
17. Kuo BC (1963) Analysis and synthesis of sampled-data control systems
18. Ziegler JG, Nichols NB (1942) Optimum settings for automatic controllers. Trans ASME 64(11)
19. Åström KJ, Hägglund T (2004) Revisiting the Ziegler-Nichols step response method for PID control. J Process Control 14(6):635–650
20. Hägglund T (2019) The one-third rule for PI controller tuning. Comput Chem Eng 127:25–30
21. Skogestad S (2003) Simple analytic rules for model reduction and PID controller tuning. J Process Control 13(4):291–309
22. Beschi M, Dormido S, Sánchez J, Visioli A (2014) Tuning of symmetric send-on-delta proportional-integral controllers. IET Control Theory Appl 8:248–259(11)

23. McNamara OP, Atherton DP (1984) Limit cycles in non linear sampled data systems. IFAC Proc Vol 17(2):507–512
24. Kuo BC (1960) The z-transform describing function for nonlinear sampled-data control systems. Proc IRE 48(5):941–942
25. Gelb A, Van der Velde WE (1968) Multiple-input describing functions and non-linear system design. McGraw-Hill

Part II
Controller Design for EBC

Chapter 6
Tuning Procedure for PI Controllers Under RQH Sampling

6.1 Introduction

Recent attention has been drawn to Event-Based Control (EBC) for continuous systems [1], driven by its advantageous features. EBC presents improved data flow management in digital networks for distributed control systems, mitigating data dropouts, package losses, and communication delays. These benefits stem from EBC's unique data transmission policy, where information is sent only upon detecting significant changes in the system state, contrasting with the periodic transmissions in classical control loops. Notably, in networked control systems, EBC is recognized as a highly promising approach, as emphasized in [2], especially within the context of modern factory automation.

In the comprehensive study in [3], a thorough overview of the major contributions to EBC in the past two decades is presented. The findings of this study reveal a relatively limited focus on Event-Based PID compared to other control methodologies like Model-based, Predictive, or State-feedback control. Despite this, PID remains widely utilized in industrial control applications, given its prevalence in various industries. Notably, due to the derivative action's sensitivity to measurement noise, PI control is the dominant form employed today [4]. This dominance, as observed in [5], persists even in the face of advanced control strategies like MPC. The versatility of PID algorithms and their continued relevance in the Industry 4.0 era, as highlighted in [6], underscore the need to adapt them for enhanced connectivity in modern industrial systems. Integrating PID with EBC principles could spread its use among these emerging application scenarios.

In EBC loops, the event generation policy holds significant importance as it governs the creation and transmission of events that regulate the controller algorithm's execution. This aspect, as emphasized in [7], directly influences the behavior and performance of such systems. Among the widely adopted event generation techniques, those based on signal quantification stand out due to their implementation simplicity. Notably, the Send-On-Delta (SOD) sampling technique is prevalent, which transmits new events when the sampled signal exceeds a change of magnitude δ. The efficacy of

O. Miguel-Escrig, *Event-Based PID Controllers with Fixed Threshold Sampling Strategies*, Springer Theses, https://doi.org/10.1007/978-3-031-62505-3_6

this strategy has been assessed in terms of control performance and communication reduction, as documented in [8, 9].

Some of the first uses of SOD in PID loops was presented in [10], aiming to reduce CPU usage in embedded control systems while maintaining loop performance. To that end, sensor unit emits samples periodically but the control is performed only when error signal surpasses certain thresholds. Problems associated to that type of control were addressed in subsequent works like [11–14]. The original idea of this type of control has derived in multiple variations, like fractional order systems [15–18].

Utilizing principles akin to SOD sampling, [19] introduced a sampling approach termed Symmetric-Send-On-Delta (SSOD). Its key characteristics include incorporating hysteresis with the same magnitude as the quantification threshold δ and employing fixed switching levels. Researchers have explored SSOD sampling for control loop, identification of processes or other practical applications. In the system identification field, [20] proposed an identification procedure using a SSOD sampler. In another study, [21], EBC employing an SSOD sampling strategy was implemented for regulating the indoor air temperature in greenhouse production processes.

Focusing on controller tuning, in [22], a tuning method for PI controllers in SSOD sampling was presented, considering First Order Plus Time Delay (FOPTD) models. Other works like [23, 24], also present tuning methods for PI controllers in SSOD-$C(s)$ loops. Furthermore, [25] explored a unified design approach for a SSOD-PID and Smith predictor suitable for self-regulating and integral processes. The methodology employed in these publications relied on the Describing Function (DF) theory. This analysis technique incorporated classical concepts of phase and gain margins into the design of EBC. The DF, a well-established analysis tool for Wiener-Hammerstein non-linear systems introduced in the 30 s and subsequently presented in [26], has various variants, such as the Dual Input DF [27], the sampled or discrete DF [28], or the fractional order DF [29]. Numerous works discussing this method can be found in the literature, for instance, in [30, 31].

An alternative sampling strategy based on signal quantization is the Regular Quantification (RQ) sampling strategy. In contrast to SSOD event generation, RQ involves sending new data whenever the sampled signal's value is a multiple of the quantization threshold δ. A comparative analysis between SSOD and RQ strategies was outlined in [32]. The RQ sampler, lacking hysteresis, may result in bursts of events and data transmission due to measurement noise. This stands as the primary drawback of RQ when compared to SSOD. Nevertheless, introducing hysteresis, as seen in the case of SSOD, imposes higher robustness requirements on the controller to prevent limit cycle oscillations.

Considering the impact of hysteresis, a novel sampling strategy termed Regular Quantization with Hysteresis (RQH) was introduced in [33] (see Chap. 2). This approach involves quantification with fixed thresholds and variable hysteresis. Depending on the chosen parameters, intermediate cases between RQ and SSOD strategies appear. In that study, ad-hoc gain and phase margins against the induction of limit cycle oscillations are proposed, and the effects on the event generation and robustness requirements is studied and compared with RQ and SSOD sampling.

Studies with these margins of classical tuning rules evinced the necessity of specific tuning algorithms for RQH-$C(s)$ loops. Thus, this chapter introduces a tuning procedure for PI controllers within this kind of loop. The proposal considers classical robustness measures such as gain and phase margins. Additionally, it incorporates the specific robustness measures, to avoid oscillations induced by the sampler. The evaluation of the influence of high order harmonics for the proposed tuning procedure, and simulation tests for various processes models prove the validity of the proposed method.

6.2 Problem Statement

Consider the networked control loop architecture presented in previous chapters (see for instance Fig. 2.1). Blocks $C(s)$ and $G(s)$ represent the controller and process transfer functions respectively, EG block stands for event generator, which in this case is the RQH sampler, a zero-order hold is represented by the block ZOH and the effect of the network is modeled by a delay $\exp(-t_d s)$.

Remarkable signals are reference signal y_r, controlled output y, disturbance input p, error signal e and sampled and held error $\bar{e}$ that arrives to controller's input. Seminal work [19] presented this loop structured considering a SSOD sampler instead of the more general RQH studied in this chapter.

RQH sampler is mainly characterized through two parameters, quantization level δ and hysteresis h. The quantization is considered to be non-null and the hysteresis can be chosen in the range $h \in [0, \delta]$. Attending to robustness, this sampler can be characterized by a single parameter h/δ.

Regarding to the input-output relationship (x-$\bar{x}$) of the RQH sampler it can be seen, and analytically demonstrated, that by fixing $h = 0$ or $h = \delta$, RQ and SSOD samplers are obtained, showing that RQH sampler constitutes a generalization that encompasses both cases.

RQH parameter choice was addressed in Chap. 2 (also in [33]), demonstrating that RQH offers an optimal trade-off between event generation and robustness requirements against limit cycle oscillations. Following the provided guidelines a RQH sampler with a given ratio h/δ is obtained. The robustness analysis was conducted using the Describing Function (DF) theory, which stated that for this kind of loops, limit cycle oscillations are inducted in the temporal response if:

$$G_{ol}(j\omega) = -\frac{1}{\mathcal{N}}, \quad \forall\omega, \tag{6.1}$$

where $G_{ol}(s) = G_p(s)C(s)e^{-t_d s}$ is the open-loop transfer function gathering the lineal behavior of the system, and $\mathcal{N}$ is the describing function representing the non-linear behavior of the sampler. Expression for $\mathcal{N}$ was obtained in Chap. 2, but is presented in equation (6.2) for the sake of completeness.

$$\mathcal{N}(A,h) = \frac{2\delta}{A\pi}\left[\sum_{k=1}^{m}\sqrt{1-\left(\frac{\delta}{A}\left(k+\frac{h}{2\delta}-\frac{1}{2}\right)\right)^2}+\right.$$
$$\left.\sum_{k=m+1}^{2m}\sqrt{1-\left(\frac{\delta}{A}\left(2m-k-\frac{h}{2\delta}+\frac{1}{2}\right)\right)^2}\right]-j\frac{2hm\delta}{A^2\pi}, \tag{6.2}$$

With the given oscillation condition, ad-hoc robustness margins against limit cycle oscillation were obtained. These margins were inspired by classical gain and phase margins, are interpreted similarly and exhibit the same behavior. The ad-hoc margins are the gain $\gamma_{h/\delta}$ and phase $\Phi_{h/\delta}$ margins to the non-linearity, which must not be surpassed to avoid limit cycle oscillations. A depiction of these margins is provided in Fig. 2.3 for a given system and non-linerity in the Nyquist and Nichols diagrams.

Using these margins, in Chap. 2, a study of the robustness against limit cycles of classical tuning methods was carried out. It was concluded that certain tuning methods cannot guaranty stable behaviors for several configurations of processes and samplers, and, on the other hand, certain methods present extremely robust behaviors, indicating that faster controllers can be deployed. This fact evinces the necessity of a specific tuning rule that guaranties stability and loop performance.

The studied methods take into account classical robustness margins, which have been demonstrated to provide poor information about stability against limit cycles. On the other hand, tuning according solely to ad-hoc margins may produce controllers with poor performance since they are measured to non-fixed points. Therefore, the proposed tuning rule in following section takes both into account.

6.3 Tuning Procedure for RQH-PI

As aforementioned, using controllers tuned according to classical tuning methods for continuous control case may lead to the induction of limit cycle oscillations. Therefore, in this section, the tuning procedure for PI controllers in RQH-PI loops is addressed. In the tuning procedure, it will be taken into account both classical and ad-hoc robustness margins. Since these last margins are based on the DF technique, the effect of unattenuated high-order harmonics is neglected.

Remark 6.1 The proposed tuning procedure starts assuming that the RQH sampler has already been chosen. Guidelines for selecting the parameters h and δ are provided in Chap. 2 based on steady-state error (e_{ss}) and peak-to-peak amplitude of measurement noise.

The proposed tuning methods aims at minimizing the IAE (Integral of Absolute value of the Error) index of the disturbance response and fulfilling the robustness requirements according to gain and phase margins. Analytically, the tuning problem to address can be express as:

$$\begin{aligned} \underset{K_p, T_i}{\text{minimize}} \quad & IE \\ \text{subject to} \quad & \gamma_{cg} \geq \gamma_{cg_r}, \qquad \Phi_{cp} \geq \Phi_{cp_r}, \\ & \gamma_{h/\delta} \geq \gamma_{h/\delta_r}, \quad \Phi_{h/\delta} \geq \Phi_{h/\delta_r} \end{aligned} \tag{6.3}$$

being γ_{cg} and Φ_{cp} the classical gain and phase margins. In the previous expression, the sub-index r refers to the requirement.

For non-oscillatory responses IAE and IE (Integral of the Error) indexes coincide, however, using IE index for calculations is preferable since it can be obtained analytically as the inverse of the controller's integral gain $K_i = K_p/T_i$.

Remark 6.2 RQH-$C(s)$ loops are known to exhibit a steady-state error, therefore, evaluating directly the IE will result in an infinite value. That is not the case for the integral of the sampled error $I\bar{E}$, which in Appendix D.2 it is shown to be similar to its continuous counterpart.

The conditions involving the constraints of optimization problem described in (6.3) can be analytically expressed as:

$$\text{Classical gain margin} \begin{cases} \left|G(j\omega_{cg})\right| \left|C(j\omega_{cg})\right| \leq 1/\gamma_{cg_r} & (6.4a) \\ \arg(G(j\omega_{cg})) + \arg(C(j\omega_{cg})) = -\pi & (6.4b) \end{cases}$$

$$\text{Classical phase margin} \begin{cases} \left|G_{ol}(j\omega_{cp})\right| = 1 & (6.5a) \\ \arg(G_{ol}(j\omega_{cp})) \leq -\pi - \Phi_{cp_r} & (6.5b) \end{cases}$$

$$h/\delta \text{ gain margin} \begin{cases} \left|G(j\omega_{\gamma})\right| \left|C(j\omega_{\gamma})\right| \leq \dfrac{1}{\gamma_{h/\delta_r}} \left|-\dfrac{1}{\mathcal{N}(A_{\gamma})}\right| & (6.6a) \\ \arg(G_{ol}(j\omega_{\gamma})) = \arg\left(-\dfrac{1}{\mathcal{N}(A_{\gamma})}\right) & (6.6b) \end{cases}$$

$$h/\delta \text{ phase margin} \begin{cases} \left|G_{ol}(j\omega_{\Phi})\right| = \left|-\dfrac{1}{\mathcal{N}(A_{\Phi})}\right| & (6.7a) \\ \arg(G_{ol}(j\omega_{\Phi})) \leq \arg\left(-\dfrac{1}{\mathcal{N}(A_{\Phi})}\right) - \Phi_{h/\delta_r} & (6.7b) \end{cases}$$

In these expressions $G(s)$ gathers the behavior of process and network delay, i.e. $G(s) = G_p(s)e^{-t_d s}$.

Since previous expressions are defined as inequalities, there exist a whole set of controllers fulfilling them, however, the objective is to obtain the controller minimizing the IE index among that set.

The tuning procedure consists of 3 steps:

1. Define the set of PI controllers obtainable for a given process.
2. Tune each of the obtained controllers to meet the gain margin requirement, and, afterwards detune the controllers not fulfill Φ_{cp}, $\gamma_{h/\delta}$ or/and $\Phi_{h/\delta}$ requirements.
3. Finally, select among the resulting controllers the one minimizing $IE = T_i/K_p$.

These steps are detailed in the following lines:

Step 1 Firstly, the set of possible values for ω_{cg} is obtained. For a PI controller $\arg(C(j\omega)) \in \left[-\frac{\pi}{2}, 0\right]$, thus, by applying equation (6.4b), $\arg(G(j\omega_{cg})) \in \left[-\pi, -\frac{\pi}{2}\right]$. The range of ω_{cg} is then obtained from the frequency response of $G(j\omega)$, obtaining the frequencies at the boundaries $\arg(G(j\omega)) = -\pi$ and $\arg(G(j\omega)) = -\frac{\pi}{2}$. A regular grid in this range is defined for ω_{cg}. Each term will correspond to a PI controller with parameters calculated in the following step.

Step 2 For each frequency in ω_{cg}, the integral time is calculated form the PI phase expression

$$\arg(C(j\omega)) = \arctan(T_i\omega) - \frac{\pi}{2}, \tag{6.8}$$

and combined with Eq. (6.4b):

$$T_i = \frac{1}{\omega_{cg}} \tan\left(-\arg(G(j\omega_{cg})) - \frac{\pi}{2}\right). \tag{6.9}$$

The corresponding proportional gain for each case is obtained from expression (6.4a), where the modulus of a PI controller is:

$$|C(j\omega)| = K_p \frac{\sqrt{1 + (\omega T_i)^2}}{\omega T_i}, \tag{6.10}$$

resulting in:

$$K_p \leq \frac{\omega_{cg} T_i}{\gamma_{cg_r} \left|G(j\omega_{cg})\right| \sqrt{1 + (\omega_{cg} T_i)^2}}. \tag{6.11}$$

Since the goal is minimizing the IE index, it will be considered the maximum gain for K_p for each of the evaluated frequencies, that is:

$$K_p = \frac{\omega_{cg} T_i}{\gamma_{cg_r} \left|G(j\omega_{cg})\right| \sqrt{1 + (\omega_{cg} T_i)^2}}. \tag{6.12}$$

Once the classical gain margin is fulfilled the restrictions expressed in equations (6.5a) to (6.7b) are evaluated, and the controller is detuned through the reduction of the proportional gain. This reduction implies a radial shrink of the open-loop transfer function in the Nyquist plane, resulting in an improvement of all the robustness margins.
To fulfill the classical phase margin requisite, equations in (6.5) must be satisfied. The term G_{ol} in that expression involves the previously calculated K_p and T_i, and the current margin Φ_{cp} is obtained. If $\Phi_{cp} \geq \Phi_{cp_r}$, the current controller does not require detuning. In case this constraint is not fulfilled, the new frequency ω_{cp} is calculated from equation (6.5b), forcing the equality

$$\arg(G_{ol}(j\omega_{cp})) = -\pi - \Phi_{cp_r}. \tag{6.13}$$

Controller is properly detuned through the factor k, which is introduced in equation (6.5a) and is calculated as:

$$k = \frac{1}{\left|G_{ol}(j\omega_{cp})\right|}. \tag{6.14}$$

The resulting new gain is kK_p. Since $0 < k < 1$ by detuning the controllers, conditions in (6.4) are still satisfied. At this point, the current controller fulfills gain and phase margins.
The remaining constraints to check and/or fulfill are those on $\gamma_{h/\delta}$ and $\Phi_{h/\delta}$. Starting with the gain margin to the non-linearity requirement, the current margin $\gamma_{h/\delta}$ is measured. If $\gamma_{h/\delta} < \gamma_{h/\delta_r}$ a detuning must be performed. A new detuning factor k is necessary, which can be derived from the definition of the ad-hoc gain margin for a system $kG_{ol}(j\omega)$ as:

$$(\omega_\gamma, A_\gamma) = \arg\min_{(\omega,A)} \left(\frac{\left|-\frac{1}{\mathcal{N}(A)}\right|}{k|G_{ol}(j\omega)|} : \arg(G_{ol}(j\omega)) = \arg\left(-\frac{1}{\mathcal{N}(A)}\right)\right). \tag{6.15}$$

Note that k does not affect the phase ($\arg(G_{ol}(j\omega)) = \arg\left(-\frac{1}{\mathcal{N}(A)}\right)$) and also that the factor $1/k$ scales $\frac{\left|-\frac{1}{\mathcal{N}(A)}\right|}{|G_{ol}(j\omega)|}$. Therefore, the frequency ω_γ and amplitude A_γ remain unchanged. The new value for k can then be obtained from expression (6.6a):

$$k = \frac{1}{\gamma_{h/\delta_r}} \left|-\frac{1}{\mathcal{N}(A_\gamma)}\right| \frac{1}{\left|G_{ol}(j\omega_\gamma)\right|}. \tag{6.16}$$

Similarly to precedent cases, the phase margin to the non-linearity $\Phi_{h/\delta}$ is checked. If it fulfills $\Phi_{h/\delta} > \Phi_{h/\delta_r}$, no detuning is required. In case this is not fulfilled, a reduction in K_p must be performed. Consider the definition of the $\Phi_{h/\delta}$ for a system defined as $kG_{ol}(j\omega)$:

$$(\omega_\Phi, A_\Phi) = \arg\min_{(\omega,A)} \left(\arg(kG_{ol}(j\omega)) - \arg\left(\frac{-1}{\mathcal{N}(A)}\right) : \left|\frac{-1}{\mathcal{N}(A)}\right| = k|G_{ol}(j\omega)| \right). \tag{6.17}$$

It is worth noting that in the previous definition, the term $\left|-\frac{1}{\mathcal{N}(A)}\right| = k|G_{ol}(j\omega)|$ depends on k, meaning that, unlike than in the case for the gain margin to the non-linearity, ω_Φ and A_Φ depend on k. The new values for parameters ω_Φ and A_Φ in which the new phase margin will be measured, after applying the detuning factor, can be calculated as:

$$(\omega'_\Phi, A'_\Phi) = \arg\max_{(\omega,A)} \left(|G_{ol}(j\omega)| - \left|-\frac{1}{\mathcal{N}'(A)}\right| \right), \tag{6.18}$$

in which $-1/\mathcal{N}'(A) = -1/\mathcal{N}(A) \cdot 1\angle\Phi_{h/\delta_r}$. The new value for k is obtained from:

$$k = \left|-\frac{1}{\mathcal{N}'(A'_\Phi)}\right| \frac{1}{|G_{ol}(j\omega'_\Phi)|}. \tag{6.19}$$

The new detuning factor guarantees zero phase margin between kG_{ol} and $-1/\mathcal{N}'$, which is $-1/\mathcal{N}$ shifted Φ_{h/δ_r} in phase, obtaining therefore the value k to fulfill the phase margin to non-linearity constraint.

Step 3 From the preceding steps a set of controllers fulfilling all the constraints and defined by their distinct pairs (K_p, T_i) is obtained. The final controller resulting from this method is obtained by filtering among the previous set the controller that minimizes the IE index, which will be the one that maximizes the K_p/T_i ratio.

The proposed tuning procedure is summarized in Fig. C.1 in Appendix C.2. This figure compacts in a single flowchart the previously detailed *Steps*.

Remark 6.3 The wear of actuators is an important factor to consider in practical applications. This wear can be limited by reducing the magnitude of the control action bumps produced by the step changes of magnitude δ that RQH sampler introduces. Considering a PI controller, these bumps are characterized by $\delta_u = K_p\delta$. According to this expression, a maximum value for K_p can be established given the quantization δ and the admissible control action bumps as: $K_{p_{max}} \leq \delta_{u_{max}}/\delta$. This constraint must be applied to all controllers obtained in Step 2.

Remark 6.4 Network latency, modeled in Fig. 2.1 as a delay t_d, must be taken into account in the design, incorporating it to the process transfer function. As it is well-known, delays reduce loop performance and robustness, therefore, it must be considered in the process transfer function as $G(s) = G_p(s)exp(-t_d s)$ in the design procedure, in the cases this communication delay is a known parameter.

Remark 6.5 The presented tuning methodology can be extended to also encompass PID tuning. Some modifications are required, for instance, expressions (6.8) and (6.10), are reformulated as follows:

$$\arg\{C(j\omega)\} = \arctan\left(\frac{\omega T_i(NT_i/T_d+1)}{NT_i/T_d-(\omega T_i)^2(1+N)}\right) - \arctan\left(\frac{\omega T_i}{NT_i/T_d}\right) - \frac{\pi}{2}, \tag{6.20}$$

$$|C(j\omega)| = K_p\frac{\sqrt{(\omega T_i(NT_i/T_d+1))^2+(NT_i/T_d-(\omega T_i)^2(1+N))^2}}{\omega T_i\sqrt{(\omega T_i)^2+(NT_i/T_d)^2}}, \tag{6.21}$$

respectively. Besides, controller's phase range is now $\left[-\frac{\pi}{2}, \frac{\pi}{2}\right]$, changing the expressions obtained from (6.4)–(6.7). Due to the inclusion of T_d and N additional constraints are required for the design.

Nonetheless, it must be born in mind that the proposed method is sustained on DF theory, and therefore, high order harmonics must be attenuated in order for the controller to be valid. It is advised to perform an harmonic analysis, which can be addressed as in the following section.

6.4 Harmonic's Influence

The ad-hoc gain ($\gamma_{h/\delta}$) and phase ($\Phi_{h/\delta}$) margins against limit cycle oscillations, used in the tuning procedure, are grounded on the DF theory (see Chap. 2). This analysis method assumes that the linear part of the system attenuates harmonics higher than the fundamental, this is known as *the filtering hypothesis*. Not fulfilling this assumption compromises DF validity to forecast limit cycle induction. In this section, the analysis of the effect of high order harmonics on the robustness margins is addressed. The analytic procedure followed in this section can be extended to evaluate other tuning methods and controllers applied in RQH-$C(s)$ loops.

Consider a sinusoidal wave with period T_o and amplitude A at the input of the sampler, which can produce single-leveled symmetric oscillations (m=1). The commutations on the output signal are produced at the time fractions ρ_1 and ρ_2, in which the input signal value is:

$$e\left(\rho_p\frac{T_o}{2}\right) = \begin{cases} \delta\left(\frac{1}{2}+\frac{h}{2\delta}\right) & \text{for } p=1 \\ \delta\left(\frac{1}{2}-\frac{h}{2\delta}\right) & \text{for } p=2 \end{cases}, \tag{6.22}$$

and the time fractions for the given input signal are:

$$\begin{aligned}\rho_1 &= \frac{1}{\pi}\arcsin\left(\frac{\delta}{A}\left(\frac{1}{2}+\frac{h}{2\delta}\right)\right)\\ \rho_2 &= 1-\frac{1}{\pi}\arcsin\left(\frac{\delta}{A}\left(\frac{1}{2}-\frac{h}{2\delta}\right)\right)\end{aligned}. \tag{6.23}$$

The sampled signal $\bar{e}(t)$ can be developed as a Fourier series, which is then operated through the loop. Assuming for the sake of simplicity $y_r(t)=0$, the resulting expression of $e(t)$ is obtained, which, evaluating at the previous time fractions, results in:

$$\begin{aligned}e\left(\rho_1\frac{T_o}{2}\right) &= -\frac{4\delta}{\pi}\sum_{n_{odd}}^{\infty}\frac{1}{n}\left[\Re\{G_{ol}(jn\omega_o)\}\frac{1}{2}\sin(n\pi(\rho_2-\rho_1))+\Im\{G_{ol}(jn\omega_o)\}\sin^2\left(\frac{n\pi}{2}(\rho_2-\rho_1)\right)\right]\\ e\left(\rho_2\frac{T_o}{2}\right) &= -\frac{4\delta}{\pi}\sum_{n_{odd}}^{\infty}\frac{1}{n}\left[\Re\{G_{ol}(jn\omega_o)\}\frac{1}{2}\sin(n\pi(\rho_2-\rho_1))-\Im\{G_{ol}(jn\omega_o)\}\sin^2\left(\frac{n\pi}{2}(\rho_2-\rho_1)\right)\right]\end{aligned}$$

Combining with expression (6.22) and substituting the time fractions ρ_1 and ρ_2 with values in (6.23):

$$\begin{aligned}\Re\{G_{ol}(j\omega_o)\} &= -\frac{\frac{\pi}{4}+\sum\limits_{n=3,5,\ldots}^{\infty}\frac{1}{n}\Re\{G_{ol}(jn\omega_o)\}\sin(n\theta)}{\sin(\theta)},\\ \Im\{G_{ol}(j\omega_o)\} &= -\frac{\frac{\pi h}{4\delta}+\sum\limits_{n=3,5,\ldots}^{\infty}\frac{1}{n}\Im\{G_{ol}(jn\omega_o)\}(1+\cos(n\theta))}{1+\cos(\theta)},\end{aligned} \tag{6.24}$$

where:

$$\theta=\arcsin\left(\frac{\delta}{2A}\left[\left(1-\frac{h}{\delta}\right)\sqrt{1-\left[\frac{\delta}{2A}\left(1+\frac{h}{\delta}\right)\right]^2}+\left(1+\frac{h}{\delta}\right)\sqrt{1-\left[\frac{\delta}{2A}\left(1-\frac{h}{\delta}\right)\right]^2}\right]\right),$$

which depends only on amplitude and sampler characteristics.

Expressions in (6.24) present the value of real and imaginary parts of $G_{ol}(j\omega_o)$ when limit cycles of frequency ω_o are induced. This expression takes into account the effect of the high order harmonics, presented by the term $n\omega_o$. Thus, this is the expression to use for validating controllers tuned according to Sect. 6.3.

To that end, it will be estimated more accurately the point with respect to which margins $\gamma_{h/\delta}$ and $\Phi_{h/\delta}$ are measured. In expressions in (6.24) it will be considered:

$$G_{ol}(jn\omega_o)=G_{ol}(jn\omega_\gamma)\gamma_{h/\delta};\quad n=3,5,\cdots \tag{6.25}$$

to evaluate the gain margin accuracy, and:

$$G_{ol}(jn\omega_o)=G_{ol}(j\omega_\Phi)\cdot 1\angle{-\Phi_{h/\delta}};\quad n=3,5,\cdots \tag{6.26}$$

to evaluate the phase margin.

The resulting point for evaluating the gain margin will be denoted by $G_{ol,\gamma}(j\omega_o)$, and must be compared to the point $-1/\mathcal{N}(A_\gamma)$. In the same way, the point $G_{ol,\Phi}(j\omega_o)$ will be compared with $-1/\mathcal{N}(A_\Phi)$ to evaluate the phase margin.

The application of this analysis procedure is described in the following example.

Example 6.1 Consider a process defined by the following transfer function:

$$G(s) = \frac{1}{(s+1)^4}.$$

A PI controller according to AMIGO method [34] is tuned with $K_p = 0.301$ and $T_i = 2.427$. The RQH sampler is considered to be defined by the ratio $h/\delta = 2/3$.

The ad-hoc margins have been measured, obtaining $\gamma_{h/\delta} = 11.51$ dB and $\Phi_{h/\delta} = 39.61°$, at frequencies $\omega_\gamma = 0.4703$ rad/s and $\omega_\Phi = 0.1218$ rad/s, and for the ratios $\delta/A = 0.834$ and $\delta/A = 0.833$, respectively.

To test the validity of the robustness margins, it has been considered the next 3 harmonics $n = 3, 5, 7$ in equations in (6.24). Expressions (6.25) and (6.26) are used to test the gain and phase margins respectively.

The location of the new critical points, according to the harmonic analysis, has been obtained. In this particular case, the variation between $-1/\mathcal{N}(A_\gamma)$ and $G_{ol,\gamma}(j\omega_o)$ and between $-1/\mathcal{N}(A_\Phi)$ and $G_{ol,\Phi}(j\omega_o)$ is not significant. Therefore, the corrected margins will be similar to the obtained with the fundamental harmonic.

Consider now that the precedent sampler is replaced by another one characterized by the ratio $h/\delta = 1/6$. Similarly to the previous example, the robustness measures against limit cycle oscillations have been measured and the effect of the harmonics has also been evaluated, obtaining the new critical points.

For this example, the variation of the critical points implies a reduction in the margins, which is clearly reflected in the phase margin, which is reduced in 10°.

Despite the variation in the margins, limit cycle oscillations are not induced in the temporal response, allowing for a safe operation. Nevertheless, this evinces that high order harmonics influence must be evaluated due to its potentially detrimental effects regarding robustness. This is specially advisable for systems with low filtering capabilities, which hinder DF theory reliability.

Summarizing, the ad-hoc margins offer an accurate order of magnitude of the robustness against limit cycle oscillations induction. Evaluation of high order harmonic effects reveals both beneficial and detrimental cases. It is recommended the evaluation of the harmonics effect for systems with low filtering linear parts. Besides, notable variations on the margins are mostly expected for those margins obtained at low frequencies with regard to the crossover phase frequency, resulting in distant points to the inverse negative of the DF locus. See for instance in last example, margin $\Phi_{h/\delta}$ which was obtained for a relative low value ω_Φ.

6.5 Illustrative Examples

This section gathers several examples illustrating the validity of the proposed tuning method, some remarkable characteristics to be taken into account and comparatives with other tuning methods and samplers.

Example 6.2 Consider that a process is described by the following transfer function model:

$$G(s) = \frac{1}{(s+1)^4}. \tag{6.27}$$

In addition, consider the peak-to-peak amplitude of the measurement noise to be of magnitude $e_{n_{p-p}} = 0.03$ and a required maximum admissible steady-state error of $e_{ss} = 0.1$. The RQH sampler is selected according to guidelines provided in Chap. 2. The resulting parameters are $h = 0.04$ and $\delta = 0.16$.

To achieve proper closed-loop response and robustness against limit cycle induction consider the following constraints for the design:

$$\begin{array}{ll} \gamma_{cg} \geq 6\,dB & \Phi_{cp} \geq 45^\circ \\ \gamma_{h/\delta} \geq 2\,dB & \Phi_{h/\delta} \geq 15^\circ \end{array}.$$

The controller resulting of applying the proposed tuning procedure has the following parameters: $K_p = 1.05$, $T_i = 2.6$; and the exact margins have been measured:

$$\begin{array}{ll} \gamma_{cg} = 7.22\,dB & \Phi_{cp} = 45^\circ \\ \gamma_{h/\delta} = 2.91\,dB & \Phi_{h/\delta} = 18.6^\circ \end{array},$$

fulfilling the requirements.

The temporal response of the system against reference and disturbance changes has been simulated and is presented in Fig. 6.1. As it can be seen, the sampler does not induct limit cycle oscillations nor generates events due to the noise.

In case that the control action bumps must be limited, the tuning procedure must consider Remark 6.3. To showcase this, consider a maximum control action variation of $\delta_{u_{max}} = 0.05$, which was exceeded in precedent case ($\delta_u = 0.168$). The parameters for the new controller are $K_p = 0.312$ and $T_i = 1.037$ and the robustness margins:

$$\begin{array}{ll} \gamma_{cg} = 9.5126\,dB & \Phi_{cp} = 45^\circ \\ \gamma_{h/\delta} = 4.486\,dB & \Phi_{h/\delta} = 20.95^\circ \end{array}.$$

The temporal response of the system has been simulated and, in this case, the variations on the control action are reduced and, as a consequence, a smoother response is obtained.

Example 6.3 Controllers tuned for RQH-$C(s)$ loops tend to require higher overall robustness with the increase of the ratio h/δ. This can induce the misconception of

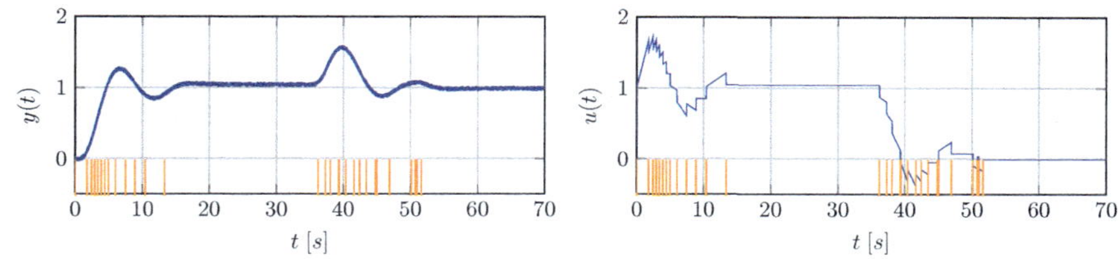

Fig. 6.1 Temporal response of the controlled output and control action to unitary step-like changes in reference and disturbance of the example system with the proposed controller and sampler. (Reprinted from Journal of the Franklin Institute, Vol 358, No. 9, Miguel-Escrig, O., & Romero-Pérez, J. A., *Tuning procedure for event-based PI controllers under regular quantization with hysteresis*, 4643-4670, Copyright (2021), with permission from Elsevier)

designing controllers for high ratios of h/δ constitute the worst case scenario, and therefore, they produce the most conservative and always valid controller regardless of the actual sampler. This example disproves this misleading statement.

Consider a process modeled by:

$$G(s) = \frac{e^{-5s}}{(s+1)^3}. \tag{6.28}$$

Besides, consider the following tuning requirements:

$$\begin{array}{ll} \gamma_{cg} \geq 6\,dB & \Phi_{cp} \geq 45^\circ \\ \gamma_{h/\delta} \geq 4\,dB & \Phi_{h/\delta} \geq 30^\circ \end{array}.$$

For the aim of this example, consider that attending to RQH parameter selection criteria, the required sampler should have a ratio $h/\delta = 1/20$. However, to disprove the precedent statement, the design is performed considering a ratio $h/\delta = 1$ (SSOD), resulting in a controller with $K_p = 0.518$ and $T_i = 5.584$. The classical gain and phase margins for this system are $\gamma_{cg} = 6.02\,dB$ and $\Phi_{cp} = 72.1^\circ$, respectively.

Regarding to the ad-hoc robustness margins, for the SSOD sampler, the obtained margins are $\gamma_{h/\delta} = 4.51\,dB$ and $\Phi_{h/\delta} = 30^\circ$, fulfilling the requirements. For the sampler characterized by the ratio $h/\delta = 1/20$ these margins are $\gamma_{h/\delta} = 3.83\,dB$ and $\Phi_{h/\delta} = 59^\circ$, not fulfilling the robustness requirements and notably reducing the gain margin to the non-linearity. Therefore, the robustness requirements are not met, disproving the original premise.

Example 6.4 In this example a comparative between SSOD-$C(s)$ and RQH-$C(s)$ loops performance, both with the proposed tuning method is presented.

Consider the setup presented in Example 6.2, keeping process, RQH sampler, noise and steady-state error requirements.

For the SSOD sampler, δ can be chosen in a range defined by the noise and steady-state error constraints:

$$e_{n_{p-p}} < \delta \leq e_{ss}.$$

Table 6.1 Controllers and performance indexes for each controller and sampler under restrictions S_1 and S_2

		K_p	T_i	$IA\bar{E}_{ref}$	$t_{s,ref}$	$IA\bar{E}_d$	$t_{s,d}$	n_{ev}
S_1	RQH	1.01	2.5	4.545	13.899	3.538	17.678	26
	SSOD	0.817	2.57	4.701	17.334	3.795	15.578	34
S_2	RQH	0.822	3.081	3.582	5.488	3.285	10.508	16
	SSOD	0.997	5.049	4.895	18.794	4.747	23.498	26

Since choosing lower values for δ increases unnecessarily the amount of generated events, $\delta = 0.1$ is selected.

Two controllers fulfilling different sets of requirements are designed. These sets, referred to as S_1 and S_2, are:

$$\begin{aligned} S_1 : \gamma_{cg} &\geq 6\,dB \quad \Phi_{cp} \geq 45^\circ & S_2 : \gamma_{cg} &\geq 6\,dB \quad \Phi_{cp} \geq 45^\circ \\ \gamma_{h/\delta} &\geq 3\,dB \quad \Phi_{h/\delta} \geq 15^\circ & \gamma_{h/\delta} &\geq 3\,dB \quad \Phi_{h/\delta} \geq 45^\circ \end{aligned}$$

Being the unique difference that set S_2 requires a higher phase margin against limit cycle oscillations than S_1.

Four controllers are obtained, two for each sampler according to the given sets of requirements, whose parameters are presented in Table 6.1.

The performance for the closed loop response is assessed through several adjusted parameters. Firstly, the Integral of the Absolute sampled Error (using $\bar{e}$ instead of e), as a consequence of unitary step changes in reference ($IA\bar{E}_{ref}$) and disturbance ($IA\bar{E}_d$) inputs. It can be expressed analytically as:

$$IA\bar{E} = \int_0^\infty |\bar{e}(t)|\,dt.$$

Besides, due to the appearance of a steady-state error, the settling time is reframed as the elapsed time from the application of the exciting signal to the cross of the last threshold stabilizing around e_{ss}. Also the number of generated events is considered.

Controlled output and control action temporal response against reference and disturbance unitary steps has been obtained through simulation.

Focusing first on the S_1 case, temporal responses for both controllers and samplers do not differ significantly. Control action for RQH sampler presents higher magnitudes and changes than for SSOD, which are produced by their respective quantization δ. Measured results from these signals presented in Table 6.1 corroborate precedent observations and reveal that, besides not degrading the performance, RQH sampling issues a lesser amount of events than SSOD to perform the closed loop control.

Table 6.2 Obtained controller and performance parameters for each tuning method

	K_p	T_i	$\bar{IAE}_{ref}$	$t_{s,ref}$	$\bar{IAE}_d$	$t_{s,d}$	n_{ev}
Proposed	0.6646	2.1206	4.8637	14.878	4.1017	13.798	41
AMIGO	0.2473	1.8807	7.4872	18.088	7.7062	22.048	31
One-Third	0.3202	2.8453	8.6168	24.088	8.6736	27.388	31
Ziegler-Nichols	0.9257	6.0792	6.3643	30.268	6.3643	34.768	43

In the S_2 case, a notable improvement of RQH performance over SSOD is observed. As in the precedent case, RQH issues less events than SSOD, but in this case a notable increase in performance is registered. This difference is justified by the slow response of SSOD controller, producing slow variations on the temporal response that worsen the overall performance.

To sum up, RQH sampling is observed to issue less events than SSOD in closed loop experiences without degrading the performance. Besides, in order to attain the same robustness than RQH-$C(s)$, controllers in SSOD-$C(s)$ loops are forced to present a slow response degrading their performance.

Example 6.5 In this example, a comparative between the proposed method and other well-known tuning methods is performed.

Consider a process modeled by:

$$G(s) = \frac{1-s}{(s+1)^3}.$$

Consider a measurement noise peak-to-peak amplitude of 0.04 units and maximum admissible steady-state error of 0.075. According to provided guidelines, RQH sampler parameters are: $h = 0.05$ and $\delta = 0.1$.

The controller tuned according to the proposed methodology is designed to meet the following constraints:

$$\begin{array}{l} \gamma_{cg} \geq 6\,dB \quad \Phi_{cp} \geq 45^\circ \\ \gamma_{h/\delta} \geq 2\,dB \quad \Phi_{h/\delta} \geq 15^\circ \end{array}.$$

The remaining controllers will be tuned according to the well-known tuning methods AMIGO [34], One-Third [35] and Ziegler-Nichols [36]. A FOPTD model approximation has been considered to apply these methods. A summary of the controllers' parameters is presented in Table 6.2.

It must be remarked that there is not a priori guarantee that AMIGO, One-Third and Ziegler-Nichols controllers avoid limit cycle oscillations, but in this case all controllers present stable behaviors.

The temporal response for these systems against unitary step changes in reference and disturbance inputs has been obtained in simulation. The resulting controlled output, control action and generated events are presented in Fig. 6.2.

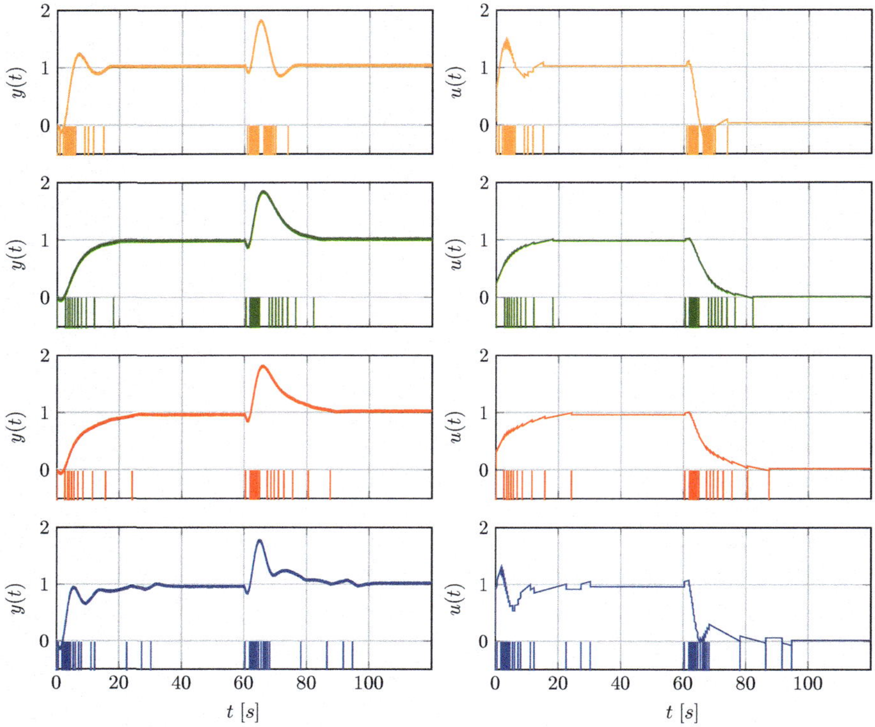

Fig. 6.2 Temporal response of the measured output of the process with different controllers, the events generated and their respective control action to unitary step changes in the reference (at $t = 0s$) and disturbance (at $t = 60s$) inputs. (Orange: proposed controller. Green: AMIGO. Red: One-Third. Blue: Ziegler-Nichols). (Reprinted from Journal of the Franklin Institute, Vol 358, No. 9, Miguel-Escrig, O., & Romero-Pérez, J. A., *Tuning procedure for event-based PI controllers under regular quantization with hysteresis*, 4643–4670, Copyright (2021), with permission from Elsevier)

All the controllers lead to stable responses, avoiding limit cycles. AMIGO and One-Third tuning rules exhibit a slower temporal response in comparison to other rules. Nevertheless, this behavior is considered normal as these rules also offer the highest level of robustness. On the other hand, the remaining controllers showcase a faster response. Despite this, there is a notable difference in the temporal response shape and settling time between Ziegler-Nichols and the proposed method.

Performance measures presented in Example 6.4 have been used to evaluate the controllers under study. Summarized data is presented in Table 6.2. The proposed method outperforms the rest in integral errors and settling times, both against reference and disturbance changes. Nevertheless, it presents a higher event generation than AMIGO and One-Third tuning rules due to their overdamped temporal response.

Quick summary:

- ☐ A tuning procedure for PI controllers in RQH-$C(s)$ loops is presented.
- ☐ The tuning procedure consists fo three steps in which classical and ad-hoc robustness margins are considered, ensuring a proper temporal response and robustness against limit cycle oscillations.
- ☐ The effect on the robustness of high order harmonics has been assessed, validating the proposed tuning method.
- ☐ The harmonic analysis can be extended to other controller structures or tuning rules. This analysis is advisable, specially, for systems with low filtering capabilities.
- ☐ The efficacy of the proposed method is illustrated through several examples. A comparison with other well-known tuning rules is provided.

References

1. Lunze J (2015) Event-based control: introduction and survey. In: Miskowicz M (ed) Event-Based control and signal processing. CRC Press, Boca Raton, pp 3–20
2. Dotoli M, Fay A, Miśkowicz M, Seatzu C (2019) An overview of current technologies and emerging trends in factory automation. Int J Prod Res 57(15–16):5047–5067
3. Aranda-Escolástico E, Guinaldo M, Heradio R, Chacón J, Vargas H, Sánchez J, Dormido S (2020) Event-based control: a bibliometric analysis of twenty years of research. IEEE Access 8(March):47188–47208
4. Åström Karl J, Tore H (2006) Advanced PID control. Systems, and automation society, ISA-the instrumentation
5. Samad T (2017) A survey on industry impact and challenges thereof. IEEE Control Syst 37(1):17–18
6. Maxim A, Copot D, Copot C, Ionescu CM (2019) The 5w's for control as part of industry 4.0: why, what, where, who, and when-A PID and MPC control perspective. Inventions 4(1)
7. Hsia TC (1974) Analytic design of adaptive sampling control law in sampled-data systems. IEEE Trans Autom Control 19(1):39–42
8. Dormido S, Sánchez J, Kofman E (2008) Muestreo, control y comunicación basados en eventos. Revista Iberoamericana de Automática e Informática Industrial RIAI 5(1):5–26
9. Ploennigs J, Vasyutynskyy V, Kabitzsch K (2010) Comparative study of energy-efficient sampling approaches for wireless control networks. IEEE Trans Ind Inf 6(3):416–424
10. Årzén K-E (1999) A simple event-based PID controller. In: Proceedings of the 14th IFAC world congress, vol 18, pp 423–428
11. Durand S, Marchand N (2009) An event-based PID controller with low computational cost. In: 8th international conference on sampling theory and applications (SampTA'09), pages Special–session
12. Durand S, Marchand N (2009) Further results on event-based PID controller. In: Control conference (ECC), 2009 European, pp 1979–1984. IEEE
13. Vasyutynskyy V, Kabitzsch K (2010) A comparative study of PID control algorithms adapted to send-on-delta sampling. In: 2010 IEEE international symposium on industrial electronics (ISIE). IEEE, pp 3373–3379

14. Vasyutynskyy V, Kabitzsch K (2009) Time constraints in PID controls with send-on-delta. IFAC Proc Vol 42(3):48–55
15. Muresan C, Birs I, Dulf EH (2020) Event-based implementation of fractional order IMC controllers for simple FOPDT processes. Mathematics 8(8):1378
16. Birs I, Muresan C, Both R, Nascu I (2020) A real life implementation of fractional order event based PI control. In: 2020 IEEE international conference on automation, quality and testing, robotics (AQTR), pp 1–6
17. Birs I, Nascu I, Ionescu C, Muresan C (2020) Event-based fractional order control. J Adv Res 25:191– 203. Recent Advances in the Fractional-Order Circuits and Systems: Theory, Design and Applications
18. Birs I, Muresan C, Ionescu C (2020) An event based implementation of a fractional order controller on a non-Newtonian transiting robot. European Control Conf 2020, ECC 2020, pp 1436–1441
19. Beschi M, Dormido S, Sánchez J, Visioli A (2012) Characterization of symmetric send-on-delta PI controllers. J Process Control 22(10):1930–1945
20. Sánchez J, Guinaldo M, Visioli A, Dormido S (2018) Identification of process transfer function parameters in event-based PI control loops. ISA Trans 75:157–171
21. Pawlowski A, Beschi M, Guzmán JL, Visioli A, Berenguel M, Dormido S (2016) Application of SSOD-PI and PI-SSOD event-based controllers to greenhouse climatic control. ISA Trans 65:525–536
22. Beschi M, Dormido S, Sánchez J, Visioli A (2014) Tuning of symmetric send-on-delta proportional-integral controllers. IET Control Theory and Appl 8:248–259(11)
23. Romero-Pérez J-A, Sanchis-Llopis R, Peñarrocha-Alós I (2014) A simple rule for tuning event-based PID controllers with symmetric send-on-delta sampling strategy. In: Proceedings of the 2014 IEEE emerging technology and factory automation (ETFA), pp 1–8, Sept 2014
24. Romero-Pérez J-A, Sanchis-Llopis R (2016) A new method for tuning PI controllers with symmetric send-on-delta sampling strategy. ISA Trans 64:161–173
25. Ruiz Á, Beschi M, Visioli A, Dormido S, Jiménez JE (2017) A unified event-based control approach for FOPTD and IPTD processes based on the filtered Smith predictor. J Frank Inst 354(2):1239–1264
26. Krylov NM, Bogoliubov NN (1949) Introduction to non-linear mechanics. Princeton University Press
27. West JC, Douce JL, Livesley RK (1956) The dual-input describing function and its use in the analysis of non-linear feedback systems. Proc IEE-Part B: Radio Electron Eng 103(10):463–473
28. Jury EI (1958) Sampled-data control systems. Wiley, New York
29. Tenreiro-Machado JA (2015) Fractional order describing functions. Signal Proc 107:389–394. Special Issue on ad hoc microphone arrays and wireless acoustic sensor networks Special Issue on Fractional Signal Processing and Applications
30. Gelb A, Van der Velde WE (1968) Multiple-input describing functions and non-linear system design. McGraw-Hill
31. Nassirharand A (2012) Computer-aided nonlinear control system design: using describing function models. Springer, London
32. Romero-Pérez J-A, Sanchis-Llopis R (2017) Tuning and robustness analysis of event-based PID controllers under different event generation strategies. Int J Control 1–38
33. Miguel-Escrig O, Romero-Pérez J-A (2020) Regular quantisation with hysteresis: a new sampling strategy for event-based PID control systems. IET control theory and applications
34. Åström KJ, Hägglund T (2004) Revisiting the Ziegler-Nichols step response method for PID control. J Process Control 14(6):635–650
35. Hägglund T (2019) The one-third rule for PI controller tuning. Comput Chem Eng 127:25–30
36. Ziegler JG, Nichols NB (1942) Optimum settings for automatic controllers. Trans ASME 64(11)

Chapter 7
Tuning of PID Controllers Under SSOD Sampling Based on Tsypkin Method

7.1 Introduction

Presently, Event-Based Control (EBC) emerges as a promising alternative to traditional time-driven control systems, aiming to diminish the required measurement frequency for control without compromising closed-loop performance. This proves essential for controllers in networked control systems, where numerous devices (sensors, actuators, controllers) share a limited bandwidth communication channel. The reduction in transmitted messages not only enhances the overall behavior of the network, preventing issues like dropouts and delays, but also aligns with the growing utilization of wireless communications in control applications. In such scenarios, minimized data transmission leads to a significant reduction in power consumption, thereby prolonging the battery life of self-powered remote sensors [1].

In Event-Based Control Systems (EBCS), the measurement frequency is significantly influenced by the strategy employed to generate events. One promising approach in this regard is the Send-On-Delta (SOD) strategy, which effectively minimizes the transmission of new data [2, 3]. The SOD method involves transmitting data from the sensor to the controller node only when the measured value exceeds a specified threshold δ [4]. A specific variant of SOD, known as Symmetric-Send-On-Delta (SSOD), is introduced in [5]. The SSOD sampler quantifies the input signal by a multiple of a constant value δ, and then transmits a new value when the input changes by δ units. A variant on this sampler, called Regular Quantization with Hysteresis (RQH), has been recently proposed in [6], proving to improve efficiency in event generation.

One crucial aspect in the analysis and design of EBCS is the presence of limit cycles, which lead to sustained oscillations in the closed-loop response. The characterization of limit cycles for various types of process models, including Integrator processes Plus Time Delay (IPTD), First Order processes Plus Time Delay (FOPTD), and Second Order processes Plus Time Delay (SOPTD), using a SSOD sampling strategy has been outlined in [7]. Understanding limit cycle appearance is essential

O. Miguel-Escrig, *Event-Based PID Controllers with Fixed Threshold Sampling Strategies*, Springer Theses, https://doi.org/10.1007/978-3-031-62505-3_7

to forecast the behavior of a system due to dynamic changes in the process. From a PID tuning perspective, a primary objective is to prevent limit cycle oscillations that could diminish the overall performance of the control system or cause excessive actuator wear.

Works like [8, 9] propose tuning methodologies for SSOD-PI loops, relying on robustness margins targeting to avoid limit cycle oscillations. To obtain these margins the DF technique was used, which allows connecting with classical control theory. A similar approach was adopted in [10], where a unified design for SSOD-PID control architecture, applicable to self-regulating and integral processes, was explored. By using methods, like the DF, that incorporate well-known concepts of classical control theory, can potentially foster the adoption in industrial applications of the event-based PID control paradigm, since these concepts are taught in undergraduate courses on automatic control.

Nonetheless, DF applicability relies on the filtering characteristics of the open-loop transfer function, which is responsible of attenuating higher order harmonics to allow the successful application of the DF. Consequently, low-order models, including FOPTD and SOPTD, commonly employed to depict real-world industrial processes, cannot be studied with this approach.

With regard to the type of controller, using a PID controller, which broadens the bandwidth compared to the PI case, can pose an additional challenge for the applicability of the DF. In [11], the authors introduced a novel robustness metric to prevent limit cycles in SSOD-PID loops, also covered in Chap. 4. This proposed measure is rooted in Tsypkin's method [12], a well-established approach for analyzing relay control systems. Consequently, the robustness metric is referred to as Tsypkin's margin (M_T). In contrast to the margins presented in [8, 9], which hinge on the DF, the M_T robustness margin is applicable not only to study systems with robust filtering characteristics but also to those characterized by low-order models and PID controllers.

In this chapter, a novel approach is introduced for tuning PID controllers designed for FOPTD systems in loops with a SSOD sampling strategy. FOPTD models admit a dimensionless representation, allowing to express possible results as functions of the delay to time constant ratio, including the tuning rules [13], which are called homogeneous. By integrating this concept with Tsypkin's margin, a comprehensive framework is established for tuning SSOD-based PID controllers. This framework can be applied to assess the performance of any homogeneous tuning rule used in this kind of loops. The general approach is subsequently applied to the tuning method introduced in [14], leading to a set of tuning rules that reach an optimal balance between the IAE index, instant control action variations, and robustness against limit cycles, which is expressed with M_T. Insights on the applicability of this tuning method to similar sampling strategies like RQH [6], presented in Chap. 2, are also provided.

7.2 Problem Statement

A networked control systems as shown in Fig. 2.1 with the specific adaptations detailed in Chap. 4 is considered. Transfer functions $C(s)$ and $G_p(s)$ correspond to controller and process respectively. Signals y_r, y, p, e and $\bar{e}$ correspond to reference, controlled output, disturbance, error and sampled error respectively. The block ZOH is a Zero-Order Hold, retaining the last value sent e^* until new data is received. Delay in the network is modeled by $\exp(-t_d s)$.

With regard to the tuning of controllers in SSOD-$C(s)$ architectures, several works have proposed algorithms mostly for PI controllers. Examples for this are [15] or [8, 9], where the DF theory was used. As demonstrated in [16] (see Chap. 4) DF analysis in this kind of loops lacks accuracy when dealing with systems with low filtering capabilities. Therefore, PID control is excluded and questions about applicability are arisen for FOPTD processes. Besides, analysis with Tsypkin method revealed that classical tuning rules for continuous control can also induce limit cycle oscillations.

These factors evince the necessity of a tuning methodology addressing this gap of knowledge. Therefore, in this chapter, the design of PID controllers, characterized by the transfer function given in expression (7.1),

$$C(s) = K_p \left(1 + \frac{1}{T_i s} + \frac{N T_d s}{T_d s + N}\right) \tag{7.1}$$

for FOPTD process models is tackled. The goal is to achieve an adequate closed-loop performance and robustness against limit cycle oscillations, inducted by the SSOD sampling strategy. To that end, Tsypkin method analysis presented in [11] and extended in Chap. 4 will be used.

7.3 SSOD-PID Controller Tuning

As aforementioned, there is a concerning lack of results regarding PID tuning in SSOD-$C(s)$ architectures. Most of the studies in literature encompass exclusively PI tuning and/or rely on techniques like the DF, whose accuracy for systems not fulfilling its filtering hypothesis has already been discussed.

Reasons for this can be deduced from studies in Sect. 4.5, which demonstrate that higher robustness margins against limit cycle oscillations through Tyspkin's margin M_T are obtained for PI than for PID controllers tuned according to well-known tuning methods. However, even if PID improvements over PI in temporal response are out of discussion, their tuning has not been yet addressed for SSOD-$C(s)$.

The proposed approach focuses on First Order Plus Time Delay (FOPTD) models, which are extensively used for approximating actual industry processes. Thus, not only PID tuning is addressed but also limitations of DF theory are covered.

FOPTD models admit a dimensionless representation, which allows generalizing results. The proposed tuning method relies on this feature. Some preliminary comments with regard to the dimensionless study are presented in the following lines.

7.3.1 Preliminary Comments on Dimensionless Analysis

Firstly, the implications of dimensionless analysis on the robustness must be analyzed. Consider process transfer function $G(s)$ to correspond with a FOPTD model where the network delay t_d is integrated in the term L:

$$G(s) = \frac{Ke^{-Ls}}{\tau s + 1}.$$

The dimensionless representation of $G(s)$ results from the transformation $\bar{s} = Ls$ and $G(\bar{s})/K$:

$$G(s) = \frac{Ke^{-Ls}}{\tau s + 1} \overset{s=\frac{\bar{s}}{L}}{\Longrightarrow} G(\bar{s}) = \frac{Ke^{-\bar{s}}}{\frac{\tau}{L}\bar{s} + 1} \overset{G(\bar{s})/K}{\Longrightarrow} \bar{G}(\bar{s}) = \frac{e^{-\bar{s}}}{\frac{\tau}{L}\bar{s} + 1}. \tag{7.2}$$

Dimensionless analysis allows expressing the results in terms of the ratio L/τ, which can be then scaled. Tuning rules admitting this dimensionless representation, called *homogeneous tuning rules*, are expressed for PID controllers according to:

$$K_p = K^{-1}\phi_1\left(\frac{L}{\tau}\right) = K^{-1}\overline{K}_p,\ T_i = L\phi_2\left(\frac{L}{\tau}\right) = L\overline{T}_i,\ T_d = L\phi_3\left(\frac{L}{\tau}\right) = L\overline{T}_d \tag{7.3}$$

Introducing previous expressions in the PID controller expression from equation (7.1):

$$C(s) = K_p\left(1 + \frac{1}{T_i s} + \frac{T_d s}{\frac{T_d s}{N} + 1}\right) \overset{s=\frac{\bar{s}}{L}}{\Longrightarrow} C(\bar{s}) = K_p\left(1 + \frac{1}{T_i/L\bar{s}} + \frac{T_d/L\bar{s}}{\frac{T_d/L\bar{s}}{N} + 1}\right)$$

$$\bar{C}(\bar{s}) = KC(\bar{s});\quad \bar{C}(\bar{s}) = \phi_1\left(1 + \frac{1}{\phi_2\bar{s}} + \frac{\phi_3\bar{s}}{\frac{\phi_3\bar{s}}{N} + 1}\right), \tag{7.4}$$

obtaining the dimensionless controller $\bar{C}(\bar{s})$. Similarly, the dimensionless open-loop transfer function can be obtained from expressions (7.2) and (7.4) as:

$$\overline{G_{ol}}(\bar{s}) = \bar{C}(\bar{s})\bar{G}(\bar{s}) = \phi_1 \left(1 + \frac{1}{\phi_2 \bar{s}} + \frac{\phi_3 \bar{s}}{\frac{\phi_3 \bar{s}}{N} + 1}\right) \frac{e^{-\bar{s}}}{\frac{\tau}{L}\bar{s} + 1}.$$

From precedent expressions, the correspondence between the dimensionless and dimensional open-loop transfer function can be derived:

$$\overline{G_{ol}}(\bar{s}) = \bar{C}(\bar{s})\bar{G}(\bar{s}) = (KC(\bar{s}))\left(\frac{G(\bar{s})}{K}\right) = C(\bar{s})G(\bar{s}) = G_{ol}(\bar{s}) = G_{ol}(sL). \tag{7.5}$$

Using homogeneous tuning rules simplifies the analysis also with regard to the Tsypkin margin M_T. The expression for M_T in SSOD-$C(s)$, given in Chap. 4 (see expression (4.12)), depends exclusively on the open-loop transfer function, which according to expression (7.5) $\overline{G_{ol}}(\bar{s}) = G_{ol}(sL)$. Therefore, the Tsypkin margin calculated for $\overline{G_{ol}}(\bar{s})$, denoted by $\overline{M_T}$, fulfills the following:

Property 7.1 Homogeneous tuning rules applied to FOPTD models can be analyzed through Tsypkin Margin in a dimensionless approach resulting in the same value for M_T in both cases:

$$\overline{M_T}(\overline{\omega_{min}}, \rho_{min}) = M_T(\omega_{min} L, \rho_{min}), \tag{7.6}$$

being the unique difference the critical frequency at which the margin is measured. However, it can be obtained from the dimensionless study $\omega_{min} = \overline{\omega_{min}}/L$.

Another important aspect to consider in the controller design for SSOD-$C(s)$ loops are the control action bumps. Abrupt changes may harm the actuator or cause malfunctioning.

These changes in the control action, denoted by δ_u, are produced by a change of magnitude δ on the sampled error signal $\bar{e}$, according to the expression:

$$\delta_u = K_p(1 + N)\delta = C(\infty)\delta, \tag{7.7}$$

being K_p the controller gain and N the derivative filter coefficient. Common practice among industry practitioners is to set N to a high value, however, researches like [17, 18] highlight its influence in the noise amplification to closed-loop performance trade-off. In SSOD-$C(s)$ loops similar behaviors are reported in [19].

By using homogeneous tuning rules the controller instant amplification $C(\infty)$ must be adapted to the dimensionless analysis as:

$$C(\infty) = K_p(1 + N) = \frac{\phi_1}{K}(1 + N) = \frac{\overline{C}(\infty)}{K}. \tag{7.8}$$

Table 7.1 Summary of characteristics for the studied systems

	PI				PID (N = 10)			
	M_T	IAE_p	IAE_r	$C(\infty)$	M_T	IAE_p	IAE_r	$C(\infty)$
$\overline{G}(\overline{s})$	0.38	3.3168	3.3458	0.9	0.0927	1.6626	1.9092	13.2
$G_{1,2}(s)$	0.38	6.6335	6.6916	0.9	0.0981	3.253	3.8184	13.2
$G_{2,1}(s)$	0.38	6.6335	3.3458	0.45	0.0927	3.253	1.9092	6.6
$G_{2,2}(s)$	0.38	13.2671	6.6916	0.45	0.0981	6.6505	3.8184	6.6

Finally, to assess the performance in the dimensionless approach, a relation between the IAE index for the system with dimensions and the dimensionless index has been extracted (see Appendix D.1):

$$IAE = |K|L \cdot \overline{IAE}. \tag{7.9}$$

Precedent properties and analysis results are illustrated in the following example.

Example 7.1 Consider the following FOPTD model transfer functions:

$$G_{1,2}(s) = \frac{e^{-2s}}{2s+1}, \quad G_{2,1}(s) = \frac{2e^{-s}}{s+1}, \quad G_{2,2}(s) = \frac{2e^{-2s}}{2s+1},$$

whose dimensionless model is

$$\overline{G}(\overline{s}) = \frac{e^{-\overline{s}}}{\overline{s}+1}$$

for all of them.

A PI and a PID controller are tuned according to Ziegler-Nichols [20] tuning rule, since it is a homogeneous tuning rule. For all the combinations between process and controller, Tsypkin's margin M_T, disturbance and reference IAE indexes (IAE_p and IAE_r respectively) and $C(\infty)$ is calculated or obtained through simulation. These parameters are gathered in Table 7.1.

As it can be seen, the results are consistent with expressions (7.6), (7.8) and (7.9). Minor discrepancies are derived from computational inaccuracies.

7.3.2 *Tuning Procedure*

As seen in the previous section, dimensionless modeling allows generalizing analysis for FOPTD models and homogeneous tuning rules through the single parameter L/τ. This dimensionless approach is followed in this section for a range of models $\overline{G}(\overline{s})$, for which controllers aiming to optimize the trade-off between performance through $\overline{IAE}$ index, smoothness of the control action through $\overline{C}(\infty)$ and robustness against limit cycle oscillations induction through M_T are calculated. Obtained results can then be extended to any system with dimensions $G(s)$ with the same L/τ ratio through equations (7.3), (7.8) and (7.9).

The proposed tuning procedure is grounded on SRB tuning algorithm [14]. This method minimizes the disturbance IAE index subject to constraints on the phase and gain margins. It is specially useful for the aim of this chapter due to its flexibility in the definition of the robustness constraints, expressed in terms of gain and phase margins. Intuitively, classical and ad-hoc robustness (in terms of M_T) should be approximately tied, therefore, this allows obtaining controllers with different degrees of robustness. Secondly, the derivative filtering coefficient N is a user defined parameter, offering the possibility of obtaining controllers with different values for $C(\infty)$. Finally, attending to results in [13], methods based on the minimization of indexes like the disturbance IAE that depends on plant sensitivity, grants that the resulting tuning rule is homogeneous, allowing generalization of the results.

Remark 7.1 SRB tuning method was originally conceived for continuous-time control. The extension of this method to SSOD-$C(s)$ is possible due to the inclusion of ad-hoc constraints, and should not be applied without checking these constraints. Other methods could be used as a base for the proposed tuning method, but SRB offers more flexibility for the intended goals.

The proposed tuning method must minimize the disturbance IAE index while producing a robust response against limit cycle oscillations induction and a smooth control action. Formally, this can be written as:

$$\begin{aligned} \underset{K_p, T_i, T_d, N}{\text{minimize}} \quad & IAE = \int_0^{\infty} |e(t)| dt \\ \text{subject to} \quad & M_T \geq M_{T_r}, \\ & C(\infty) \leq C_r(\infty) \end{aligned} \tag{7.10}$$

where M_{T_r} and $C_r(\infty)$ are the required Tsypkin margin and noise amplification, respectively.

Since the Tsypkin margin is not measured from a fixed point, its relationship with the design parameters of SRB method, which are the classical gain and phase margins (ϕ_m, γ_m) and the derivative filtering coefficient (N), is not evident.

Therefore, it is proposed to obtain controllers using certain grids on SRB design parameters. The characteristics of this set of controllers is then registered in terms of M_T, $\overline{IAE}$, $\overline{C}(\infty)$ and their dimensionless parameters $\overline{K_p}$, $\overline{T_i}$, $\overline{T_d}$ are stored. Afterwards, the controller fulfilling the optimization problem in (7.10) is selected. Since sets of controllers are obtained for a batch of $\overline{G}(\overline{s})$, its results are extensible for other dimensional FOPTD models with the same ratio L/τ.

7.3.3 Calculating the Dimensionless Controllers

According to the described approach, controller's dimensionless parameters using the SRB tuning method are calculated for a grid of processes $\overline{G}(\overline{s})$ with $L/\tau \in]0, 3]$. For each of those processes, SRB method is applied with all the combinations of the following design parameters grid: $\phi_m = [50°, 52.5°, 55°, 57.5°, 60°, 62.5°, 65°, 67.5°, 70°]$ and $N = [0, 0.5, 1, 2, 3, 4, 5, 6, 7, 8, 9, 10, 12, 14, 15, 16, 18, 20]$. The gain margin requirement is set to $\gamma_m \geq 6$ dB.

It is worth noting that selected grids in ϕ_m and N are small enough to ensure that the obtained controller is as close as possible to the exact controller minimizing the IAE index as stated in Eq. (7.10). The obtained sets of controllers are valid for any dimensional FOPTD model with $L/\tau \in]0, 3]$.

Remark 7.2 Alike to the stated in Property 7.1, classical phase and gain margins are maintained from dimensionless models to the dimensional ones. As in the case of M_T it only exists a scaling in the frequency at which those margins are measured.

Remark 7.3 Due to the proposed tuning procedure characteristics, any controller resulting from this method will fulfill not only the required M_{T_r} and $C_r(\infty)$, but also the classical margins specifications $\phi_m \geq 50°$ and $\gamma_m \geq 6$ dB.

The evolution of the Tsypkin margin M_T with the ratio L/τ for systems composed of dimensionless FOPTD processes and SRB controllers tuned with certain values of ϕ_m and N has been obtained. Results show that for all cases there is a process with a critical ratio L/τ from which lower values of this ratio will always produce oscillatory responses ($M_T = 0$). Tsypkin's margin is then observed to increase and then stabilize to a certain values as L/τ increases.

Besides, results show certain correspondence between phase margin and Tsypkin margin, as ϕ_m increases so does M_T. Another relationship is observed between the design parameter N and the robustness against limit cycle induction, by increasing N the robustness in terms of M_T is reduced. This trend is shown only for certain range of N since there is no significant difference between robustness for $N = 10$, 15 or 20.

$\overline{IAE}$ and $\overline{C}(\infty)$ have also been obtained. For the $\overline{IAE}$ index the behavior is as expected, PI controllers ($N = 0$) present the worst index, which is improved as N increases. Besides, phase margin ϕ_m effect on $\overline{IAE}$ index for PID controllers is not remarkable, but it is for PI controllers.

Similarly, the expected behavior is observed in $\overline{C}(\infty)$, which increases with N. Varying the design parameter ϕ_m has limited effect on $\overline{C}(\infty)$.

Remark 7.4 From the previous results, a smooth relationship is observed between robustness or performance indexes (M_T, $\overline{IAE}$ and $\overline{C}(\infty)$) and design parameters (ϕ_m and N). This suggests that the proposed tuning methodology is not very sensitive to variations on these parameters, and, thus, the proposed solution from pre-calculated sets of controllers will approximate correctly the optimization problem in (7.10).

7.3.4 Software Tool for Tuning

From the calculations described in the previous section, a set of 162 controllers, defined by $\overline{K_p}$, $\overline{T_i}$ and $\overline{T_d}$, have been calculated for each considered values of $L/\tau \in]0, 3]$, one for each value of the grid conformed by ϕ_m and N. For each of them, their respective M_T, $\overline{IAE}$ and $\overline{C}(\infty)$ have been stored. To manage the stored data, a software tool has been developed to facilitate the controller selection.

The tuning procedure has been implemented in a Java application containing the sets of pre-calculated controllers. The required M_{T_r} and $C_r(\infty)$ are introduced together with the extracted dimensionless FOPTD model parameters. With these values, the controllers fulfilling the requirements are selected and the one with minimum $\overline{IAE}$ is chosen. The dimensional parameters K_p, T_i and T_d are then retrieved.

In case that arbitrary processes not defined by a FOPTD model are introduced, the application estimates a FOPTD model approximation and resumes the tuning procedure. The developed application can be downloaded freely from: https://sites.google.com/a/uji.es/freepidtools/ssodTsypkinPid.

An illustrative example using this application is then provided.

Example 7.2 For a better visualization of the presented results, the reader is encouraged to download the application and follow the example.

Consider a process modeled by the following transfer function:

$$G(s) = \frac{1}{(s + \frac{1}{3})(s + 1)(s + 3)}.$$

Firstly, the process model is introduced in the `Plant_definition` tab. The application internally calculates the FOPTD approximation needed for tuning the controller.

Then, the design parameters are introduced in the `PID_design` tab. For this example, $M_{T_r} \geq 0.1$ and $C_r(\infty) \leq 20$ have been considered. Once the requirements are introduced the application retrieves the controller according to the optimization problem in Eq. (7.10) by pushing the button below.

The frequency response of the controller with the original system, not the FOPTD approximation, together with other design results are shown in this tab. It can be seen that the robustness margin for the original process is $M_T = 0.32$, and $C(\infty) = 18.28$, fulfilling both the requirements. The frequency response in the Nyquist diagram also displays the critical Tsypkin band and, if desired, a set of bands for other frequencies.

Another window can be opened to display the temporal response of the controlled system to a unitary step change in the reference and disturbance inputs. The magnitude of the quantization δ can also be selected.

As a result of applying the tuning procedure, not only PID controllers can be obtained, but also PI controllers (considered before as PID with $N = 0$). Consider now, for the same process, another set of constraints defined by $M_{T_r} \geq 0.1$ and

$C_r(\infty) \leq 2$. The changes are introduced in the corresponding tab. For this case, the resulting controller fulfilling the restrictions is a PI. Due to the change to a PI, the disturbance IAE index has risen, but a proper response is still obtained.

7.3.5 Tuning Table

Alternatively to the developed Java application, a table is also provided to perform the tuning. The table is presented in Appendix C.1 for several requirements on $\overline{C}(\infty)$ and $M_{T_r} > 0.2$. Obviously, the options with the table are more limited than with the application.

In the table are gathered the dimensionless parameters $\overline{K_p}$, $\overline{T_i}$ and N for processes with different L/τ ratios. $\overline{T_d}$ is omitted since it can be calculated as $\overline{T_d} = \overline{T_i}/4$. Empty entries on the table mean that there are no controllers fulfilling the requirements from the pre-calculated batch. For entries marked with $*$ the controller fulfilling the precedent $\overline{C}(\infty)$ must be taken.

The following example illustrate tuning using the provided table.

Example 7.3 Consider a process modeled by the following FOPTD model:

$$G(s) = \frac{0.5e^{-2s}}{5s+1}.$$

In the tuning table, the requirement $M_{T_r} > 0.2$ is set by default.

The FOPTD model of the process is characterized by a L/τ ratio of 0.4. According to Table C.1, 4 controllers are obtained for each of the preset requirements on $\overline{C}(\infty)$. The controllers are numbered from left to right on the table and will be referred to as C_i. Since Table C.1 presents the parameters in a dimensionless fashion, these given parameters must be converted using expression (7.3).

The resulting open-loop transfer function for all the systems, together with its critical Tsypkin branch has been obtained in the Nyquist plane. From obtained results, the constraint on M_T is fulfilled in all cases. The temporal closed-loop response of these systems against a unitary step change in the reference and disturbance inputs has been obtained through simulation. As expected, stable responses are obtained.

Differences in the resulting response are, nonetheless, observable. Controllers with greater values of $C(\infty)$, which correspond to those with low IAE, present faster responses and abrupt control action changes, produced by greater values of N. This change is most noticeable comparing C_1, which is a PI controller, with the rest of controllers, that are PID.

For this example, choosing high values of $\overline{C}(\infty)$ does not lead to a significant decrease of the IAE index, therefore, it is not justified the rise in the control action bumps. For instance, controllers C_3 and C_4, present similar temporal responses but a significantly different δ_u: $\delta_u = K_p(1+N)\delta = 3.74(1+4)0.1 = 1.87$ for C_3, and $\delta_u = 8.694$ for C_4.

7.4 Effect of the FOPTD Approximation

The presented tuning method is conceived for processes defined by a FOPTD model. Nonetheless, actual industrial processes are defined by diverse transfer function models. If the proposed method is to be applied, these processes must be approximated by a FOPTD model. The effect of this approximation on the final performance must be evaluated, specially regarding to the robustness against limit cycle oscillations.

To address this study, it has been considered a batch of 95 process models (7.11) reflecting most of the behaviors found in real applications [21].

$$
\begin{aligned}
&G(s) = \frac{e^{-s}}{(Ts+1)^2},\\
&T = 0.3, 0.5, 0.7, 1, 1.3, 1.5, 2, 4, 6, 8, 10, 20, 50, 100, 200, 500\\
&G(s) = \frac{1}{(s+1)(Ts+1)^2},\\
&T = 0.05, 0.1, 0.2, 0.5, 2, 5, 10\\
&G(s) = \frac{1}{(s+1)^n},\\
&n = 3, 4, 5, 6, 7, 8\\
&G(s) = \frac{1}{(s+1)(\alpha s+1)(\alpha^2 s+1)(\alpha^3 s+1)},\\
&\alpha = 0.1, 0.2, 0.3, 0.4, 0.5, 0.6, 0.7, 0.8, 0.9\\
&G(s) = \frac{Te^{-L_1 s}}{(T_1 s+1)(Ts+1)}, \quad T_1 + L_1 = 1,\\
&T = 1, 2, 5, 10, \quad L_1 = 0.01, 0.02, 0.05, 0.1, 0.3, 0.5, 0.7, 0.9, 1\\
&G(s) = \frac{1-\alpha s}{(s+1)^3},\\
&\alpha = 0.1, 0.2, 0.3, 0.4, 0.5, 0.6, 0.7, 0.8, 0.9, 1, 1.1\\
&G(s) = \frac{1}{(s+1)((sT)^2+1.4sT+1)},\\
&T = 0.1, 0.2, 0.3, 0.4, 0.5, 0.6, 0.7, 0.8, 0.9, 1
\end{aligned}
\tag{7.11}
$$

For each of these processes its FOPTD approximation has been computed and a controller with $M_T \geq 0.1$ and any $C(\infty)$ has been obtained. The Tsypkin margin M_T has been calculated for both, the original process and the controller, and the FOPTD approximation and the controller. The results are presented in Fig. 7.1. For all the studied cases, the value of M_T is greater for the original process than for the FOPTD approximation.

Thus, robustness against limit cycle oscillations is assured for non-FOPTD processes by applying the proposed method to their FOPTD model approximation.

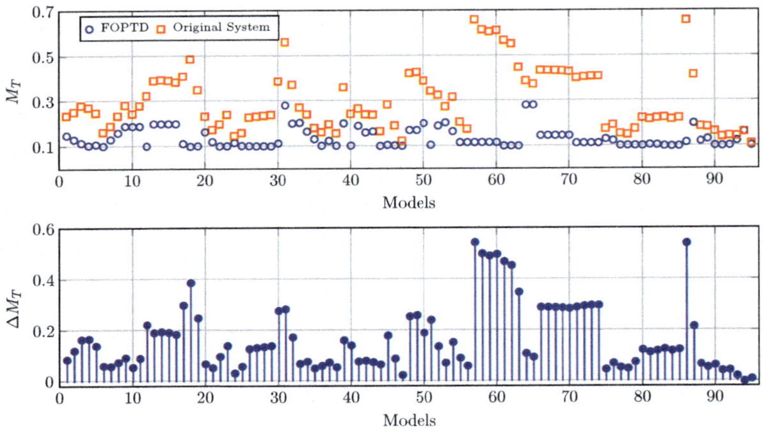

Fig. 7.1 M_T and ΔM_T for the considered systems. (Reprinted from Journal of the Franklin Institute, Vol 357, No. 2, Miguel-Escrig, O., Romero-Pérez, J. A., & Sanchis-Llopis, R., *Tuning PID controllers with symmetric send-on-delta sampling strategy*, 832–862, Copyright (2020), with permission from Elsevier)

7.5 Applicability to RQH-$C(s)$ Loops

Regular Quantization with Hysteresis has been presented and analyzed in several chapters of this book, and a tuning procedure has been proposed for RQH-PI loops based on DF theory (see Chap. 6).

Analysis of other well-known tuning rules through DF theory indicate that the adapted gain and phase margins to limit cycle induction due to the sampling tend to increase as the ratio h/δ decreases (see Chap. 2).

Besides, a comprehensive study was carried out in Chap. 4 using the Tsypkin harmonic analysis. This study evaluated the effect of high-order harmonics in the limit cycle oscillations inducted by RQH sampler. It was determined that a variant of the Tsypkin margin M_T with the same properties as for the SSOD case is applicable.

With regard to the effect of reducing the ratio h/δ, analysis with the Tsypkin margin pointed in the same line that DF based analysis.

Conjecture 7.1 Due to the observed evidence in the analysis through different techniques, the proposed tuning method in this chapter for SSOD-$C(s)$ loops, may also be applied to RQH-$C(s)$ loops.

In this sense, consider the following example.

Example 7.4 Consider a process described by the FOPTD model presented in Example 4.1 (FOPTD with $K = 1$ and $L/\tau = 0.2$), and a controller tuned according to SRB method, also presented in that example ($K_p = 2.181$, $T_i = 0.484$, $T_d = 0.115$, $N = 10$). Several ratios h/δ for the RQH are chosen: $h/\delta = [1, 3/4, 1/2, 1/4]$, note that SSOD is also included.

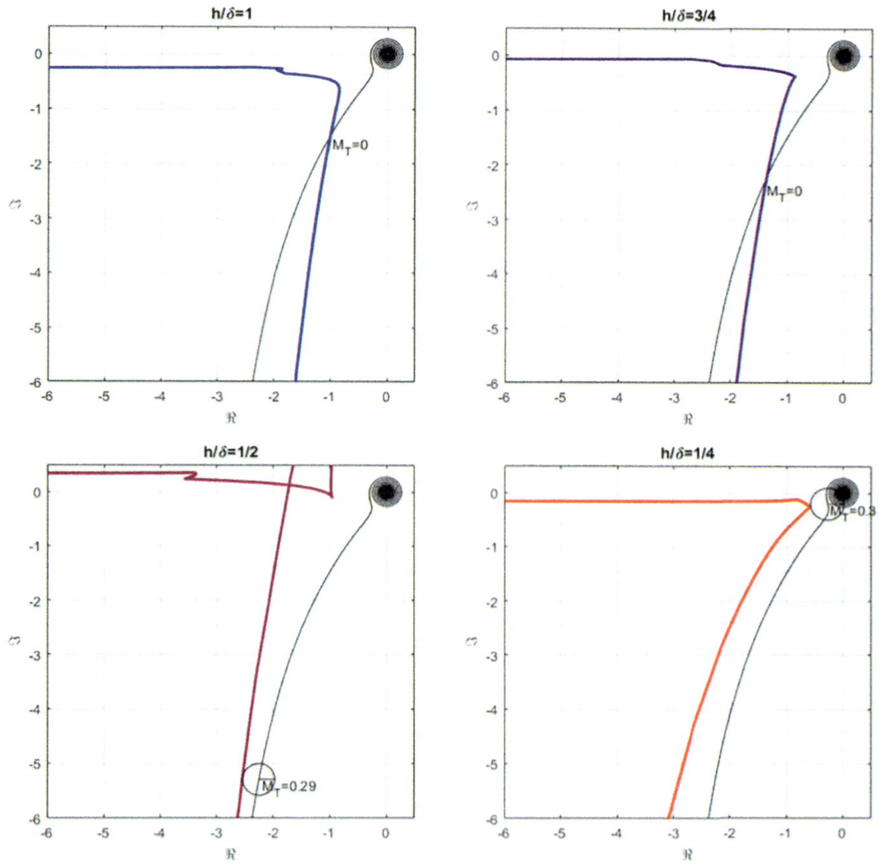

Fig. 7.2 Nyquist plot of the open-loop transfer function with all the controllers and their minimum Tsypkin branch

In Example 4.1 it was shown that under a SSOD sampling, limit cycle oscillations were induced. The Tsypkin margin M_T has been obtained for each of the presented cases (according to expression (4.25)). The critical Tsypkin branch and the open-loop transfer function for each of the studied cases is presented in Fig. 7.2. As it can be seen, $h/\delta = 1$ and $h/\delta = 3/4$ cases present $M_T = 0$, which indicates the induction of limit cycle oscillations. For the rest of cases a positive M_T is obtained, avoiding oscillations.

The temporal response of these systems against a disturbance input change of magnitude 2.5 has been simulated, and it is presented in Fig. 7.3. The analysis results are corroborated.

This results point in the same direction that Conjecture 7.1, showing that an oscillatory behavior is avoided by reducing the ratio h/δ.

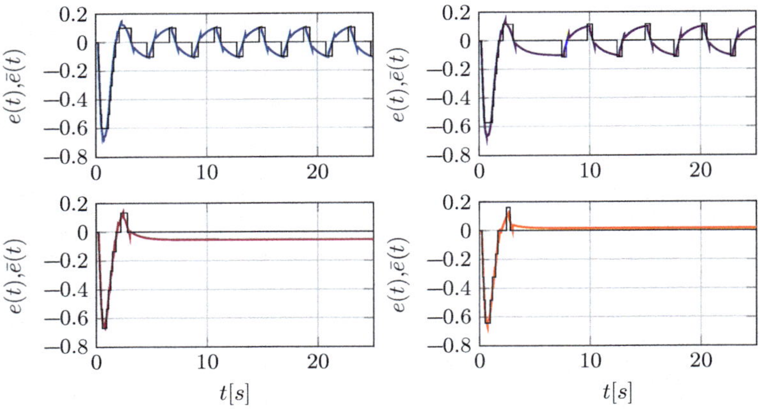

Fig. 7.3 Temporal response of the error signal $e(t)$ and sampled error signal $\bar{e}(t)$ against disturbance step change of magnitude 2.5 for the four samplers. (Same disposition and colors as in Fig. 7.2)

Remark 7.5 While several aforementioned results point towards the indicated in Conjecture 7.1, there are also some counterexamples (through the DF analysis) provided in this book, for instance, Example 6.3. Therefore, it is advised to check the robustness margins for each specific case using RQH, specially, for systems with low filtering capabilities and low M_{T_r} requirement.

Quick summary:

- ☐ A tuning method for FOPTD models in SSOD-$C(s)$ loops is presented.
- ☐ The resulting controllers provide robust responses while minimizing the IAE index and control action bumps.
- ☐ Dimensionless analysis allowed obtaining the set of optimal controllers. Derivative filter N prove crucial for achieving customized controllers with balanced performance.
- ☐ For applying the tuning method a tuning table and a freely available Java tool is provided.
- ☐ The tuning method is demonstrated to be applicable to non-FOPTD processes by approximating their temporal response with a FOPTD model.
- ☐ Discussion on the applicability of the proposed tuning method for RQH-$C(s)$ loops is provided. Specific analysis for certain cases is recommended.

References

1. Feeney L-M, Nilsson M (2001) Investigating the energy consumption of a wireless network interface in an ad hoc networking environment. In: Proceedings IEEE INFOCOM 2001. Conference on computer communications. Twentieth annual joint conference of the IEEE computer and communications society (Cat. No. 01CH37213), vol 3, pp 1548 –1557
2. Dormido S, Sánchez J, Kofman E (2008) Muestreo, control y comunicación basados en eventos. Revista Iberoamericana de Automática e Informática Industrial RIAI 5(1):5–26
3. Ploennigs J, Vasyutynskyy V, Kabitzsch K (2010) Comparative study of energy-efficient sampling approaches for wireless control networks. IEEE Trans Ind Inf 6(3):416 –424
4. Miskowicz M (2006) Send-on-delta concept: an event-based data reporting strategy. Sensors 6(1):49–63
5. Beschi M, Dormido S, Sánchez J, Visioli A (2012) Characterization of symmetric send-on-delta PI controllers. J Process Control 22(10):1930–1945
6. Miguel-Escrig O, Romero-Pérez J-A (2020) Regular quantisation with hysteresis: a new sampling strategy for event-based PID control systems. IET Control Theory Appl
7. Chacón J, Sánchez J, Visioli A, Yebra L, Dormido S (2013) Characterization of limit cycles for self-regulating and integral processes with PI control and send-on-delta sampling. J Process Control 23(6):826–838
8. Romero-Pérez J-A, Sanchis-Llopis R, Peñarrocha-Alós I (2014) A simple rule for tuning event-based PID controllers with symmetric send-on-delta sampling strategy. In: Proceedings of the 2014 IEEE emerging technology and factory automation (ETFA), pp 1–8, Sept 2014
9. Romero-Pérez J-A, Sanchis-Llopis R (2016) A new method for tuning PI controllers with symmetric send-on-delta sampling strategy. ISA Trans 64:161–173
10. Ruiz Á, Beschi M, Visioli A, Dormido S, Jiménez JE (2017) A unified event-based control approach for FOPTD and IPTD processes based on the filtered Smith predictor. J Franklin Inst 354(2):1239–1264
11. Miguel-Escrig O, Romero-Pérez J-A, Sanchis-Llopis R (2018) New robustness measure for a kind of event-based PID. IFAC-PapersOnLine 51(4):781–786
12. Tsypkin YZ (1984) Relay control systems. Cambridge University Press
13. Balaguer-Herrero P (2013) Application of dimensional analysis in systems modeling and control design. The Institution of Engineering and Technology
14. Sanchis-Llopis R, Romero-Pérez J-A, Balaguer-Herrero P (2010) Tuning of PID controllers based on simplified single parameter optimisation. Int J Control 83(9):1785–1798
15. Beschi M, Dormido S, Sánchez J, Visioli A (2014) Tuning of symmetric send-on-delta proportional-integral controllers. IET Control Theory Appl 8:248–259(11)
16. Miguel-Escrig O, Romero-Pérez J-A, Sanchis-Llopis R (2020) Tuning PID controllers with symmetric send-on-delta sampling strategy. J Franklin Inst 357(2):832–862
17. Graebe SF, Isaksson AJ (2002) Derivative filter is an integral part of PID design. IEE Proc Control Theory Appl 149(1):41–45
18. Kristiansson B, Lennartson B (2006) Robust tuning of PI and PID controllers. Using derivative action despite the sensor noise. IEEE Control Syst Mag (February):55–68
19. Romero-Pérez J-A, Sanchis-Llopis R (2017) Tuning and robustness analysis of event-based PID controllers under different event generation strategies. Int J Control 1–38
20. Ziegler JG, Nichols NB (1942) Optimum settings for automatic controllers. Trans ASME 64(11)
21. Åström KJ, Hägglund T (2006) Advanced PID Control. ISA-The Instrumentation Systems, and Automation Society

Chapter 8
Implementation and Experimental Evaluation of SSOD and RQH Sampling Strategies for EBC

8.1 Introduction

During the last decade numerous investigations have been carried out about Event-Based Control (EBC) of continuous systems. The EBC pursuits two main objectives: (1) reducing the amount of information needed to perform the closed-loop control and (2) decreasing the mean computational cost that control algorithms require.

In Event-Based Controllers, the control algorithm is only executed after the occurrence of asynchronous events that indicate a significant change in the system state, whereas in the conventional time-based controllers, whose use is more extended, the control algorithm is executed periodically. Some examples of asynchronous event generation are given in [1], among them, one of the easiest to implement is the generation of events whenever the error signal surpasses certain thresholds.

Among the earliest contributions to event based PID control it can be found the seminal paper of Årzén [2], where the main objective was to reduce the CPU usage of the control systems without significantly disturbing the performance of the control loop. Some posterior works have been focused on resolving some issues of the initial proposal, mainly, related with the calculus of the integral time by Durand [3, 4] and Vasyutynskyy [5, 6].

In his work, Årzén proposes a periodic call to the control algorithm, not being periodic its execution, i.e. the control action is updated only when needed. The algorithm includes an event generation logic based on the difference between the actual error and the error used for the last actualization of the control action. If this difference is greater than a given threshold, then the controller output is recalculated. Moreover, the time without refreshing the control action is also limited.

The event generation technique used by Årzén relative to the error signal is known as Send-On-Delta (SOD) and it has been widely used because of its simplicity. This technique has derived into more complex rules like the Symmetric-Send-On-Delta (SSOD) [7], which essentially maintains the same functioning that the SOD but adds an hysteresis of value Delta and it has fixed thresholds. Other works as [8] study

O. Miguel-Escrig, *Event-Based PID Controllers with Fixed Threshold Sampling Strategies*, Springer Theses, https://doi.org/10.1007/978-3-031-62505-3_8

the benefits of reducing the magnitude of this hysteresis, resulting in a non-linearity called Regular Quantification with Hysteresis (RQH), which presents benefits in the number of events generated while reducing some robustness requirements.

Several studies have been conducted to propose some tuning methods that take into account the effect of the SSOD sampling on the overall performance of the loop. Some works like [9] follow a more experimental approach based on classical control performance measures, others such as [10–12] use some robustness measures defined specifically for SSOD to perform the tuning procedure. These robustness measures or the techniques used to develop them have also been used to evaluate the performance of RQH sampling [8, 13].

However, in the literature aforementioned, the effects caused by the implementation of such control loops on their performance is not studied deeply. Recently, some advances have been made in relation with event-based data acquisition systems [14]. The samplers that implement event generation techniques similar to the required by Årzén's algorithm, as that presented in [15], are very complex. Besides, most of the target industrial control devices that can implement EBC and benefit from its advantages usually operate in a periodic fashion, due to the standard in which they are implemented IEC 61131 [16], contradicting the asynchronous nature of EBC.

Nevertheless, other implementation options are available like IEC 61499 standard [17], which has been developed for programming distributed control systems. This standard introduces new concepts in comparison to its predecessor, mainly, whereas the IEC 61131 is based on scan cycles, the IEC 61499, whose goal is to favor the design of distributed and reconfigurable control applications, is based on the management and treatment of events. Hence, as the application frame of IEC 61499 standard and EBC is shared and the fields of study that are currently been studied complement each other, their combination will result in a symbiotic relationship between both areas.

Therefore, the standard IEC 61499 will be used to develop the implementation of EBC using SSOD and RQH sampling strategies. Then, with the obtained implementation of EBC, a distributed application in standard IEC 61499 will be configured to evaluate experimentally the performance and the robustness of several controller configurations presented in precedent chapters. To that end, several measures will be used to compare them and asses quantitatively their performance.

8.2 Principles of IEC 61499 Programming Standard

It is important to remark that standard IEC 61499 appears as an answer to the necessity of implementing distributed applications in a standardized framework with ease, and not as a new programming methodology. Therefore, the focus of this standard revolves around the possibility of implementing applications deployed in several devices. The standard comprises several views or models of the system, each of them focusing on modeling different aspects.

The standard IEC 61499 is settled around the concept of Function Block (FB), which is the minimal functional unit of software with its own data structure that can be manipulated by one or more algorithms. In FBs interfaces coalesce data, which can have any type defined in IEC 61131, e.g. int or boolean, and events which regulate the execution control of the FB. These two types of data appear at the input and output interface of the FB. FBs are organized in types, each type having the same data structure, and can exist several instances of each type.

Depending on how FBs are formally defined, i.e. how the internal behavior is presented, they belong to different categories:

- A **Basic Function Block** is defined using a state-transition diagram called Execution Control Chart (ECC), in which algorithms are triggered as response to input events and output data and events are triggered as a consequence.
- **Composite Function Blocks** are defined by a network of FB instances. The behavior will depend on the FBs used and how are they connected.
- **Service Interface Function Blocks** permit to interact with external services from the interface of a FB. For example, they allow the communication between devices or reading the value of the inputs of a device.

FBs are gathered in a model that defines an IEC 61499 application, called Application Model. An application is defined in terms of behavior and structure by the instances of FB used and the interconnection of events and data between them, being this setup sufficient to regulate the behavior of the application. In this model the general behavior of the whole application is presented, regardless of the devices that will run the application. The FBs that implement some technical details, but they are not strictly part of the application behavior are not present in this model, e.g. FBs that allow communication between devices.

As it has been said, applications in this standard can be deployed in several devices and/or resources. The model that describes the devices, their resources, the networks and the connection between each of these elements is the System Model. This model describes the infrastructure of the system and allows assigning some properties, e.g. assigning IP addresses to devices for an Ethernet connection.

Once the elements are defined, each of the FBs that appears in the application model can be mapped to its device, assigning in this way who will be in charge of executing each element. To include the FBs necessary for example for the communication between devices that have been omitted in the Application Model, the standard provides an additional model called Device Model. In this model, the part of the application belonging to each device is presented, and here the additional FBs supporting the correct execution of the application are added.

The standard provides other models describing different aspects of the execution or the application, but for the scope of this work those models will not be explained in detail here.

8.3 Development of RQH FB in IEC 61499

One of the key aspects of standard IEC 61499 is the re-usability of the code. The development of applications is a proof of that because they are defined in terms of function blocks, which can be instantiated multiple times in different applications. Therefore, the RQH FB will not be designed to take into account device-specific properties, and it will only contain the quantization. Therefore, a Basic Function Block will be used.

The input-output relationship of the RQH quantization is defined in Eq. (2.1). From here the data that belong to the interface of the FB are extracted, being the signal $x(t)$, the quantization level δ and the hysteresis h inputs to de FB, and $\bar{x}(t)$ its output. For notation issues the output of the quantization has been renamed as y. Note that the parameter i is not listed among the input-output data, and it will be considered as an internal variable of the FB.

The input interface of the FB will be completed with an initialization event (INIT) and a request event (REQ). The INIT event will update the value of δ and h and the REQ event only the value of the input x. The output interface will be composed of an initialization event (INITO), a confirmation event (CNF) and an additional event (EVT). In addition, a boolean variable Change is added to the data interface. The appearance of the final FB RQH_quant can be seen in Fig. 8.1. The open-source program 4DIAC [18] has been used to model and develop the FBs, applications and system configuration used, and to deploy them on the physical resources.

Regarding the behavior of the FB, the values of the sampler will be taken at the initialization with INIT event, an event INITO will be sent, and the FB will remain in an Idle state waiting for requests. If while in this state another INIT event is sent, the FB will be reseted. If a sampling request (REQ) is received, the FB will compute the quantization presented in Eq. (2.1) and send a confirmation event (CNF). If a change in the output is detected, the boolean variable Change will became true and an event EVT will be sent. Either if a change is produced or not, the FB will return to its idle state. The ECC that describes this behavior is presented in Fig. 8.2, where it can be seen that the transition that evaluates if it has been produced a change in the system (true=Change) has a higher priority than the return to the Idle state, denoted by a number at the origin of the transition, being this number the order in evaluating the transitions.

Fig. 8.1 Interface of the FB RQH_quant

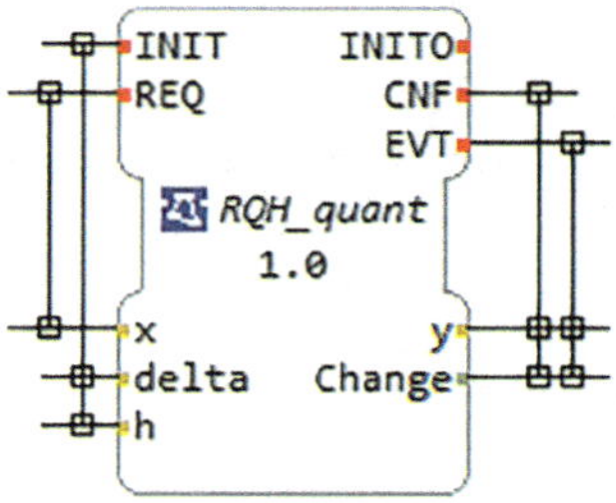

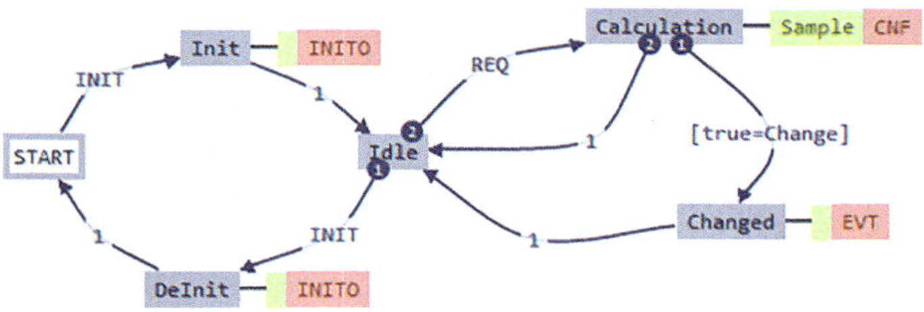

Fig. 8.2 Execution control chart describing the internal behavior of the FB RQH_quant

Fig. 8.3 Example of the functioning of the RQH_quant FB. Signal to sample in solid black, theoretic result in solid blue, output of the FB in solid red, REQ events in red arrows and EVT events in violet arrows

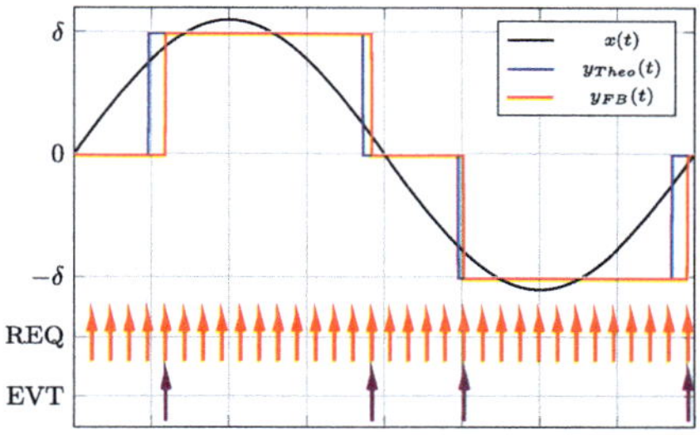

An example of the functioning of this FB is shown in Fig. 8.3, where it has been presented in black solid line the signal to sample, in blue the theoretic output of the sampler and in red the output of the FB. Additionally, REQ (which are the same that CNF) and EVT events are also indicated with red and violet arrows respectively. Depending on the request frequency different behaviors can be obtained, if the frequency is selected high enough with regard to the process dynamics, a behavior similar to a continuous sampling can be obtained. If that frequency is lowered, as shown in Fig. 8.3 a sampled non-linearity behavior as the shown in Chap. 5 will be obtained.

8.4 Study Framework

Once the FB implementing the functionality of a RQH quantifier has been developed, the control implementation can be tackled. As it has been said, in IEC 61499 programming the applications are organized and regulated by an ensemble of FBs and their connections. These FBs are mapped to a certain device, which is responsible of providing the resources necessary for their execution, and their disposition is presented in the system model. In the next lines the experimental setup and its link to these implementation concepts is described.

The experimental setup chosen to validate the theoretical results presented in previous chapters consists of a laboratory scale system composed of two connected water tanks and a pump. The tanks are placed at different heights and connected

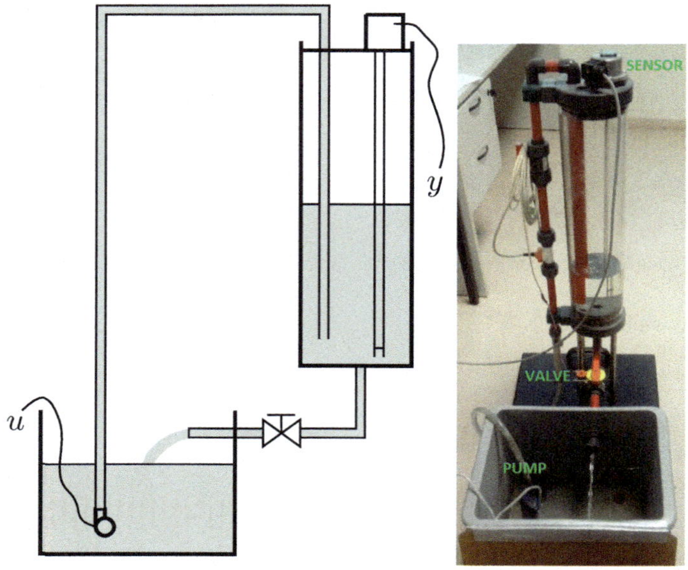

Fig. 8.4 System with two connected tanks. Schematic representation and real system

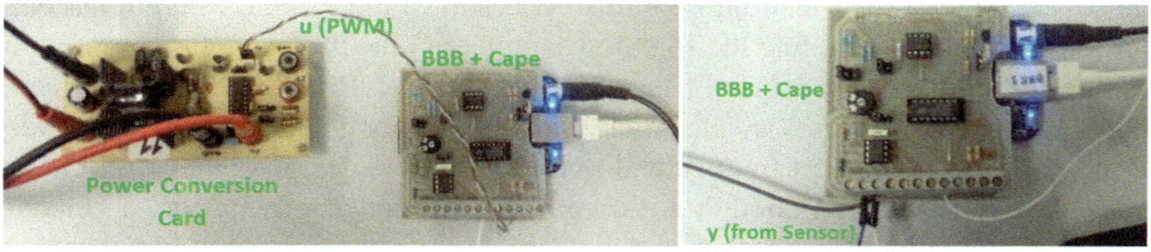

Fig. 8.5 Detail of the actuator (left) and sensor (right) units

between them with a pipe with a manual valve. The pump is placed in the lower tank and pours water into the upper tank. The goal is to control the level of liquid in the upper tank by actuating on the pump. The described system is depicted in Fig. 8.4.

The upper tank counts with a level sensor which is connected to a BeableBone Black card, which is in charge of the data acquisition, and the pump is controlled with another BeagleBone Black which has a PWM signal as control output, whose power is increased by a specific card. Varying the duty rate of the PWM the pump can be regulated. A detailed view of the sensor and actuator units is included in Fig. 8.5. Disturbances are introduced in the system by limiting the output flow with the manual valve in the pipe connecting both tanks.

These actuator and sensor units are connected to a PC that provides an user interface for displaying information about the current state of the system and allowing to modify the set-point value through an Ethernet network. The sensor and actuator

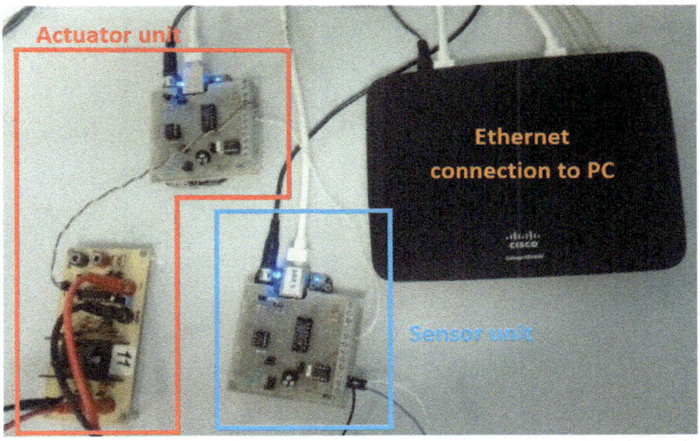

Fig. 8.6 Actuator and sensor units. The connection between elements is supported via Ethernet

units and its link to the PC, which is done via a router, are shown in Fig. 8.6. About the data flow over the network, the actuator unit receives data from the sensor unit and from the PC when a change on the set-point is introduced through the user interface, which also receives data of the measured signal and the control action applied to the process.

The presented framework is thus arranged in three different systems, namely, the sensor unit, the actuator unit and a user interface.These devices are presented in Fig. 8.7 where it can be seen that they are connected to one another through an Ethernet network. Regarding to the application, each unit will develop some additional tasks to ensure its correct functioning: (1) the sensor unit, will be in charge of sampling, quantifying, and transmitting the measured signal (2) the actuator unit, will receive the measured signal, compute the control algorithms and modify the control output signal and (3) the interface management unit will provide information about the current state of the system and will allow regulating the set-point.

The general layout that will be used for the application model is presented in Fig. 8.8, where it can be seen in a different color the FB mapped to each device. The data displayed in this figure can vary depending on the experiment being run, but the underlying ideas of a distributed control loop and how it is implemented in IEC 61499 are presented in this figure. It has to be noticed that even though the execution of the FBs entails a given execution period defined by the instance "Sampling period" of the `E_CYCLE` FB, this period has been chosen small enough (200 ms), in the cases where a continuous control is to be tested, with regard to the dynamic of the system, so the controller can be considered as a continuous one, such as it is assumed in previous chapters for its analysis.

It must be remarked that more FBs than the presented in the application model are necessary to implement a distributed application, which are modeled in the device model. These FBs are mainly communication-related FBs and their implementation is straightforward following the example tutorials in 4DIAC web page. For the concrete case under study the device models for each of the three devices are presented in Appendix E, in which these communication FBs are used to enable the communication between devices.

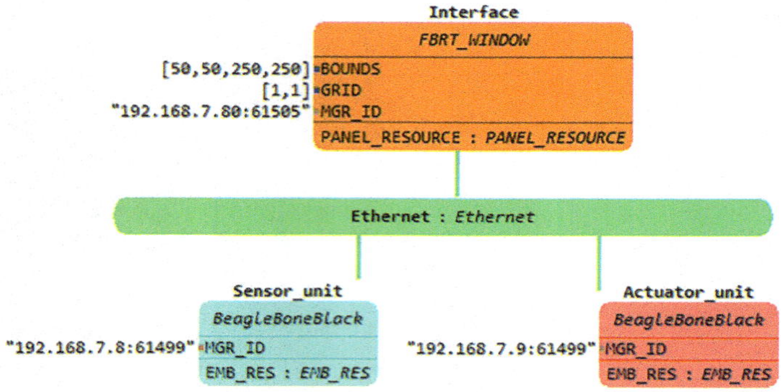

Fig. 8.7 Devices of the study framework and their connection in the system model

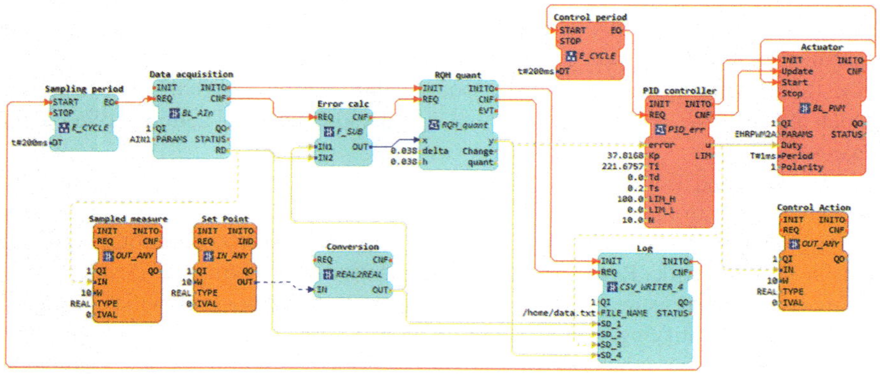

Fig. 8.8 General layout of the application model of the study framework

Additionally, in order to deploy any 4diac application the resource must contain and execute the runtime which supports the application called FORTE. Some pre-compiled versions of FORTE can be found on 4diac's web page [18] for different platforms which contain a basic set of FBs. However, as some custom FBs have been developed the runtime must be recompiled with the new additions. The cards and the PC are provided with this recompiled runtime.

8.5 Experimental Validation of EBC Controllers

Regarding the general application layout presented in Fig. 8.8 several specific configurations can be set to validate the control situations presented in Chaps. 5, 6 and 7.

The system above presented has been identified using the step response method by the following SOPTD model:

$$G(s) = \frac{0.0756e^{-2s}}{(143s + 1)(6s + 1)} \quad \left[\frac{V}{\%}\right]. \tag{8.1}$$

which relates the duty cycle of the PWM signal driving the pump and the voltage in the output of the level sensor, that is related to the level of liquid in the tank through the ratio 250 cm/9 V.

This model describes the system dynamic around the middle of the level sensor operation range, that is, in a safe region of operation where there is no risk of overflowing. The measurement noise is mostly gathered in a band of 0.01 V (0.28 cm in tank level), which is assumed as the peak-to-peak amplitude of the noise.

The experiments will be carried out in the middle of the tank, where the system has been identified, so the reference command will start at 0.5 V and will increase in 0.4 V for a step change (11 cm in the tank level). It will be considered as an acceptable steady-state error around 9.5% of this change resulting in $e_{ss} = 0.038$ V (1 cm in tank level).

For the RQH sampler design the hysteresis is chosen slightly greater than the measured noise $h = 0.012$ V (1/3 cm), and by applying Eq. (2.2) the value of the quantization level is obtained $\delta = 0.064$ V (1.8 cm), resulting a RQH ratio $h/\delta = 3/16$. For the SSOD the quantization value is equal to the hysteresis and to the maximum steady-state error $\delta = 0.038$.

To tune the controllers for these samplers, two sets of conditions are presented in terms of classical and specific margins to the non-linearity according to the tuning procedure presented in Chap. 2. One of these sets presents low robustness margins (SLR) to the non-linearity, relying the robustness of the system exclusively on the classical robustness measures. The other set (SHR), in addition to these classical robustness measures, also requires some high robustness margins to the non-linearity.

$$\begin{aligned} SLR:\ &\gamma_{cg} \geq 6\,dB \;\; \Phi_{cp} \geq 45^\circ & SHR:\ &\gamma_{cg} \geq 6\,dB \;\; \Phi_{cp} \geq 45^\circ \\ &\gamma_{h/\delta} \geq 0\,dB \;\; \Phi_{h/\delta} \geq 0^\circ & &\gamma_{h/\delta} \geq 3\,dB \;\; \Phi_{h/\delta} \geq 45^\circ \end{aligned}$$

Also, according to the approach presented in Chap. 7, a controller has been tunned using the tuning table presented in Appendix C.1. As the approximated FOPTD model has a L/τ close to 0.1, the first entry of the table has been used.

The controller's parameters using the proposed method are gathered in Table 8.1 where the parameters obtained by applying some classical tuning rules, namely Ziegler-Nichols [19] and AMIGO [20], have also been presented.

Table 8.1 Parameters for the different PI controllers for the two tank system

	SLR		SHR		AMIGO	Ziegler Nichols	Tsypkin
	SSOD	RQH	SSOD	RQH			
K_p	157.37	157.37	37.82	93.76	45.96	142.05	52.51
T_i	44.76	44.76	221.68	107.76	80.36	36.11	90.76

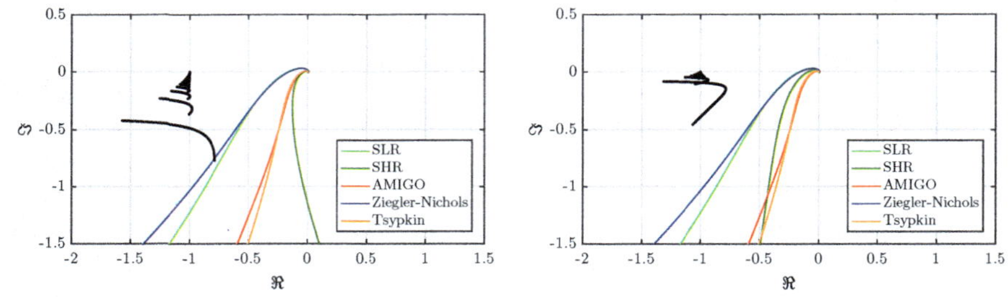

(a) SSOD (b) RQH

Fig. 8.9 Frequency response of the open-loop transfer function and the DF traces of the corresponding sampler

8.5.1 DF Method Analysis

Before testing these controllers on the real system a preliminary study based on the DF technique can be conducted for attempting to forecast the existence of limit cycle oscillations. The frequency response of the open-loop transfer function obtained with these controllers together with the negative inverse of the describing function of the studied RQH sampler can be seen in Fig. 8.9.

According to the theoretical analysis, when a RQH sampling is applied to the loop all the controllers should avoid inducing limit cycle oscillations to the system since it does not exist intersection between the open-loop transfer function and the traces of the DF. However, in the SSOD case, both Ziegler-Nichols and SLR case present a similar frequency response, and therefore it is expected a limit cycle oscillation from them. The other methods should also avoid limit cycles.

8.5.2 Experimental Results

The temporal responses of the system with different controllers under SSOD sampling can be seen in Fig. 8.10. In these figures, the response against reference change and disturbance is presented. The reference signal is represented in black, the sampled measured signal in purple, and the temporal responses in a different color for each controller: blue for Ziegler-Nichols, red for AMIGO, light green for SLR, dark green for SHR and orange for Tsypkin.

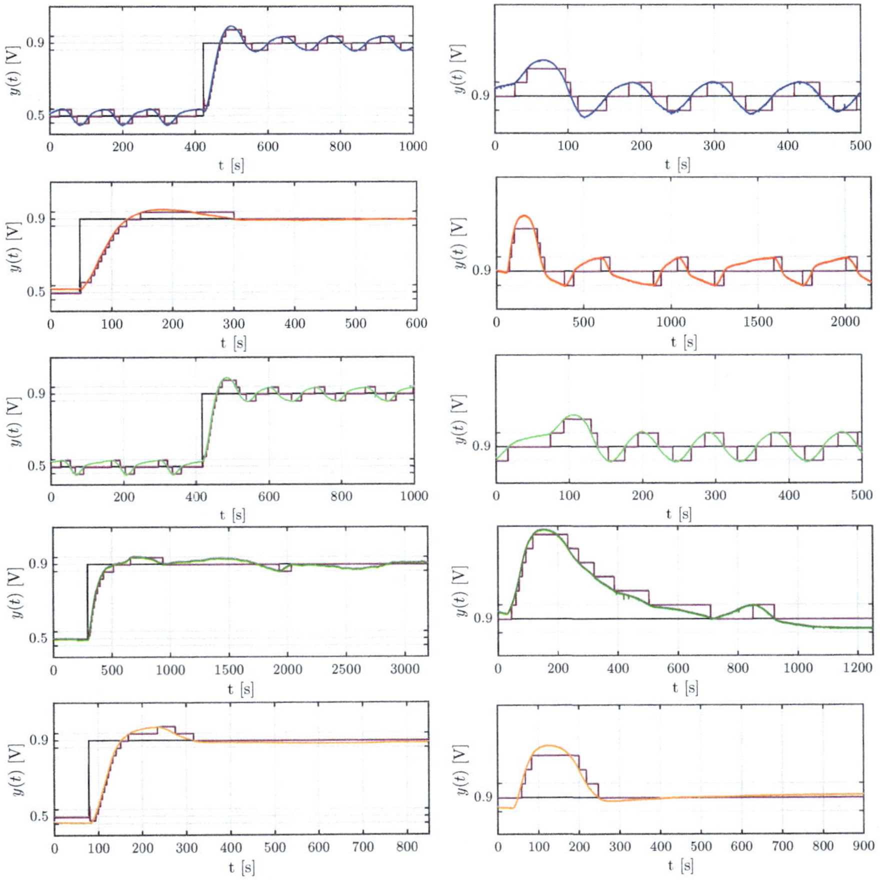

Fig. 8.10 Temporal response of the system with different controllers under SSOD sampling. (Blue: Ziegler-Nichols, Red: AMIGO, Light green: SLR, Dark green: SHR, Orange: Tsypkin)

From these figures it can be seen that Ziegler-Nichols and SLR controllers induce limit cycle oscillations in the system, which was expected since the traces of their respective open-loop transfer function intersect or are close to intersecting the DF traces. Regarding the temporal responses obtained with AMIGO controller it can be seen that it does not conduce to a limit cycle on the step change response. However, regarding the disturbance change the system enters a limit cycle. This limit cycle is not predicted by the DF technique since the oscillation type is very different from the assumed sinusoidal-like oscillations at the input of the non-linearity, therefore, further analysis is required to determine the robustness provided by this controller, which will be addressed in the next section. The proposed SHR and Tsypkin controllers avoid limit cycle oscillations as expected from the theoretical analysis.

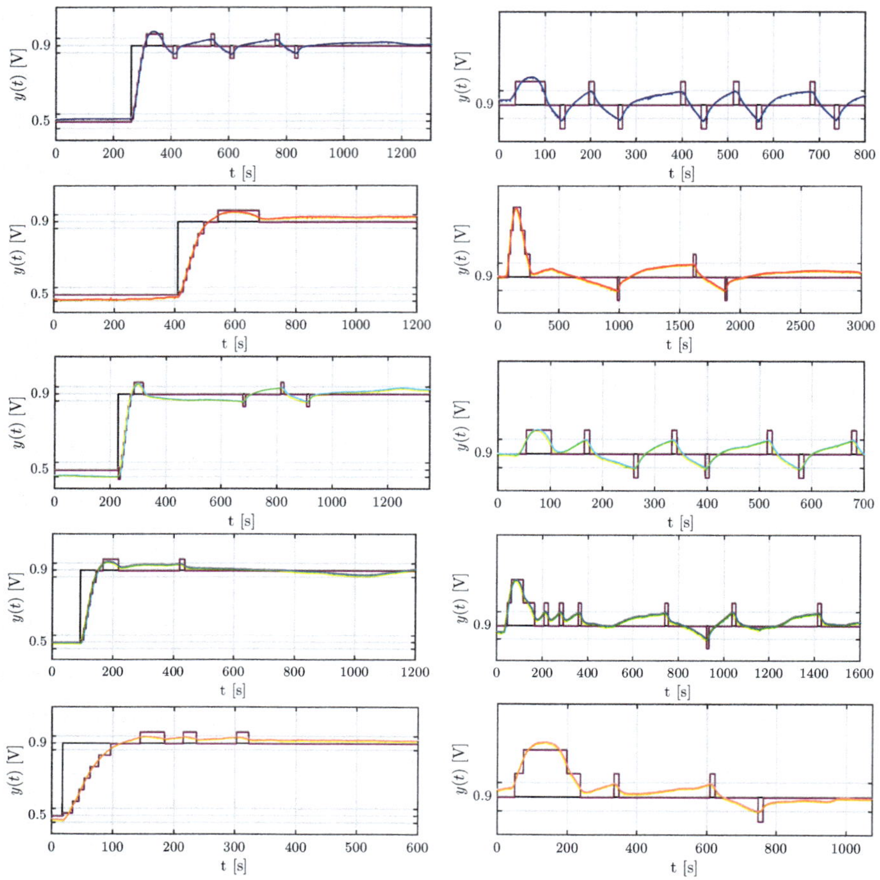

Fig. 8.11 Temporal response of the system with different controllers under RQH sampling. (Blue: Ziegler-Nichols, Red: AMIGO, Light green: SLR, Dark green: SHR, Orange: Tsypkin)

The temporal responses obtained when replacing the SSOD by a RQH sampling are presented in Fig. 8.11. From the theoretical study with the DF technique none of these controllers should induce the system to an oscillatory state. However, Ziegler-Nichols and SLR controllers still induce a limit cycle oscillation in their disturbance temporal response. This fact does not match the theoretical result based on the DF since their open-loop transfer function does not intersect the DF traces and they present a phase margin to the non-linearity of $\Phi_{h/\delta} = 21°$ for Ziegler-Nichols and $\Phi_{h/\delta} = 24°$ for SLR. However, it is true that the oscillations are very different from the sinusoidal input at the non-linearity that the DF technique assumes, and therefore, this kind of oscillations cannot be predicted with the DF technique. Hence, these cases will be studied with the Tsypkin Margin.

Regarding AMIGO's, SHR's and Tsypkin's step and disturbance responses it can be concluded that they do not induce limit cycle oscillations in the system response. It is true that there exist some additional switches, specially in SHR response against disturbance, but are produced as a response to variations in the system.

8.5.3 Tsypkin's Method Analysis

The precedent temporal responses present both oscillatory and non-oscillatory behaviors. Some of those behaviors were correctly predicted by the DF method, however, in those cases where hardly sinusoidal oscillations are present the DF fails to predict the apparition of a limit cycle oscillation.

Regarding the SSOD case it was expected from Ziegler-Nichols and SLR controllers to induce limit cycle oscillations in the response and from SHR and Tsypkin controllers to avoid them. In those cases the prediction provided by the DF technique is correct. This can be seen in Fig. 8.12, where the open-loop transfer function for each of the described cases has been presented with their critical Tsypkin Branch (purple) which defines the robustness measure M_T. In those figures the traces of the inverse negative of the DF (black) have been also presented, which in the four cases the trace $m = 1$ is very similar to the critical Tsypkin Branch (which always represents oscillations of $m = 1$ type), showing the validity of the DF method in those

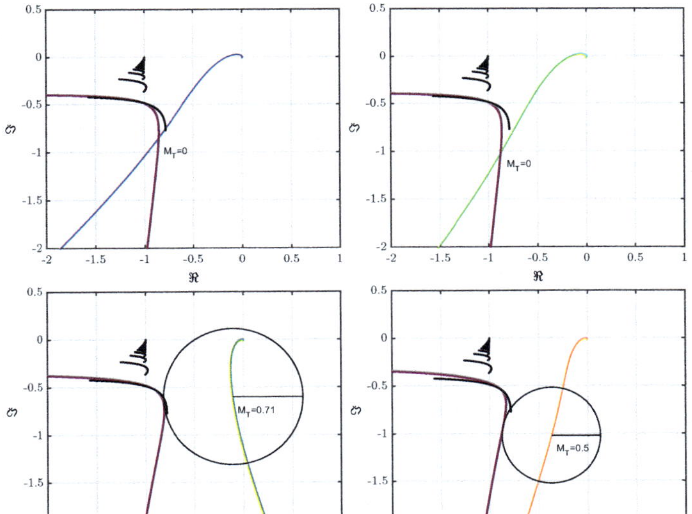

Fig. 8.12 Critical Tsypkin Branch and M_T for those cases where the DF is safely applicable. (Blue: Ziegler-Nichols, Light green: SLR, Dark green: SHR, Orange: Tsypkin)

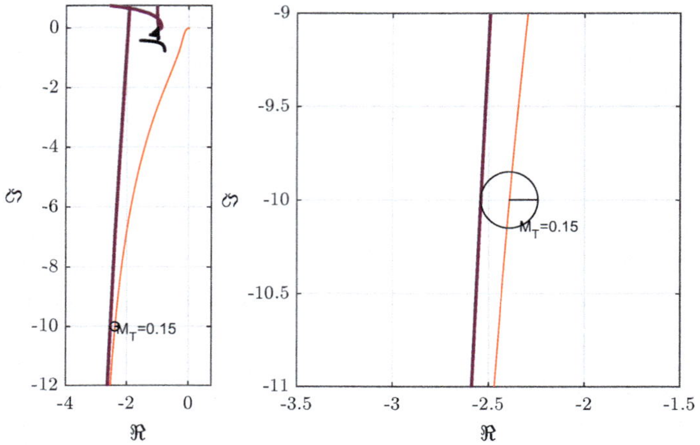

Fig. 8.13 Critical Tsypkin Branch and M_T for AMIGO controller. Whole representation and detail

cases. Also in this figure it can be seen how by applying the Tsypkin method the limit cycles are predicted, for Ziegler-Nichols and SLR controllers the resulting M_T is equal to 0 which indicates the apparition of a limit cycle. On the other hand, for SHR the robustness measure is $M_T = 0.71$ and for Tsypkin's controller $M_T = 0.5$, measures that widely avoid the apparition of limit cycle oscillations. It is remarkable the increase of robustness from Tsypkin controller, the parameters selected from the tuning table assured a robustness $M_T \geq 0.2$ for the FOPTD model approximation, which is increased up to $M_T = 0.5$ for the SOPTD model.

However, the AMIGO controller arose questions about its applicability since it largely avoided the intersection with the DF traces, but in the experimental response to disturbances it presented a behavior very similar to an oscillation where there are unattenuated harmonics, therefore, an analysis with the Tsypkin method is necessary.

As in the precedent cases studied, the open-loop transfer function have been presented in Fig. 8.13 with its critical Tsypkin Branch and the obtained M_T. In the left image of the presented figure it can be noticed that the critical value of M_T is obtained at a very low frequency compared to where oscillations are predicted with the Describing Function. Hence, it is not surprising that the critical Tsypkin Branch does not resemble the DF traces. The value of robustness obtained $M_T = 0.15$ should be enough to ensure that no limit cycle oscillations will be present in the loop. However, it must be pointed out that this value of M_T has been obtained for very low frequency values with regard to the crossover frequency. This is important since the Tsypkin method requires to have well characterized the transfer function of the process under study. The identification method used in this experimental part, the step response method, provides several solutions that fit the experimental temporal data without being necessarily accurate in the frequency response. That is to say that both a FOPTD model and a second order model may provide an acceptable identification result despite being their frequency responses very different.

Therefore, since the variations that produce the uncertainties in the identification procedure (or even a possible change of model to identify with) in the frequency response, specially in the low frequencies, may be significant, it is recommended to be careful of designs that ensure robustness at low frequencies if the frequency response is not well characterized.

In this concrete example, the usage of AMIGO's controller is not advisable for the reasons aforementioned. However, with a proper frequency response identification AMIGO tuning method usually provides robust controllers.

Specially interesting is the Tsypkin analysis in the RQH case, where despite the fact that the open-loop transfer functions do not intersect the traces of the DF, some controllers still induced limit cycle oscillations in the temporal responses. Nevertheless, these oscillations are hardly sinusoidal, having the high order harmonics a remarkable influence on them.

The Tsypkin analysis has been carried out for all the controllers involved in the RQH-PI experiment and the obtained results are shown in Fig. 8.14. In this figure

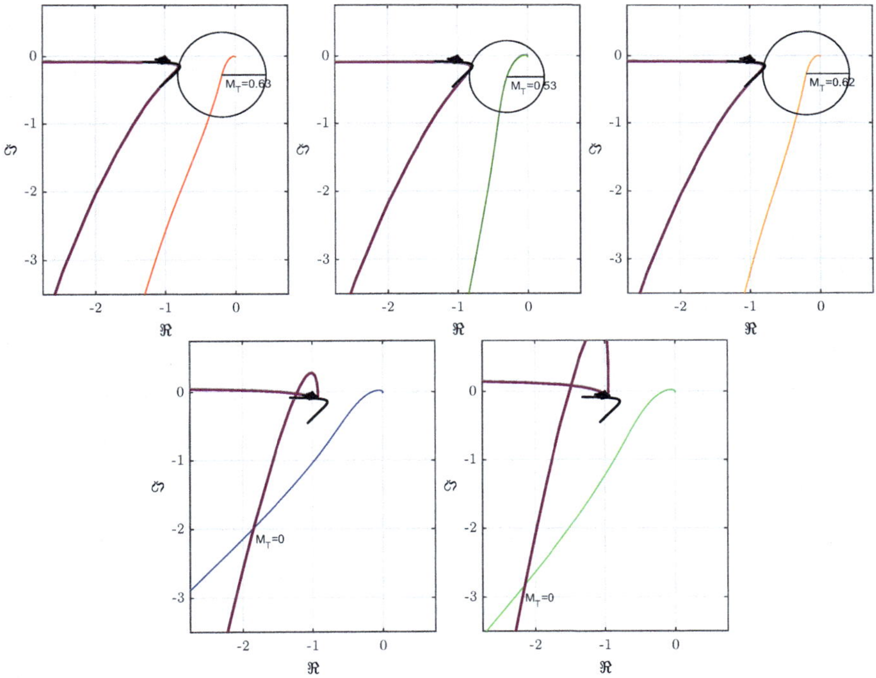

Fig. 8.14 Critical Tsypkin Branch and M_T for RQH-PI loops (Red: AMIGO, Dark green: SHR, Orange: Tsypkin, Blue: Ziegler-Nichols, Light green: SLR)

the open-loop transfer function of each case has been represented in the Nyquist diagram with the critical Tsypkin branch in purple. In addition, the DF traces have been presented in black.

The analysis reveals that Ziegler-Nichols and SLR controllers can present limit cycle oscillations and that AMIGO, SHR and Tsypkin controllers avoid them, which corresponds with the experimental temporal responses shown in Fig. 8.11.

It is also remarkable the similarities and differences between the critical Tsypkin branch and the DF traces. In those cases where unpredicted oscillations occur the DF traces differ significantly from the Tsypkin branch that leads to $M_T = 0$, which is expected since the oscillations that appear are very different from the assumed sinusoid at the input of the non-linearity. However, in the other cases the critical Tsypkin branch corresponds and extends the trace $m = 1$ of the DF approach.

8.5.4 Control Period Influence

The analysis performed in previous sections is based on all linear elements being continuous. However, the controller is discrete since it has been implemented in a BeagleBoneBlack card. This discrete implementation of the controller is predominant in any practical application and the periodicity in which the algorithms are executed could play a significant role. In previous experiments, the period of the control algorithm has been chosen very small with regard to the dynamics of the process, in that way, the analysis remains valid and its effect on the final results is negligible.

To evaluate the effect of the control period it will be considered the SSOD case with the SHR controller for which the DF approach is valid as it is shown in Fig. 8.12 where it can be seen how the DF trace matches the critical Tsypin Branch, and has proven to have the largest robustness margin against limit cycle oscillations. One criteria to choose the control period is by evaluating the step response of the system. A common choice is to select it from 20 to 10 times smaller than the rise time $T_s \in [T_r/20, T_r/10]$. The rise time for the closed-loop step response of the theoretical process with SHR controller is $T_r = 141$ s and in the previous sections the control period has been $T_s = 200$ ms.

Knowing that the implementation of the controller corresponds to a Backward Euler implementation of the integral term, the Sampled DF method has been applied to this system. The Nyquist diagram with the normalized open-loop transfer function $G_{ol}(s)/T_s = C^*(s)G(s)ZOH(s)/T_s$ and the convex hulls encircling the sampled DF traces for several values of r (samples per oscillation period) with their respective robustness measure is presented in Fig. 8.15a. In this figure, it can be seen that the convex hulls are very similar to the continuous DF traces and therefore that the impact of the control period is negligible. In addition, the robustness against limit cycle oscillations has been measured to be $D_{ch} = 0.69$ for $r = 978$, which is very similar to the obtained M_T for the continuous case.

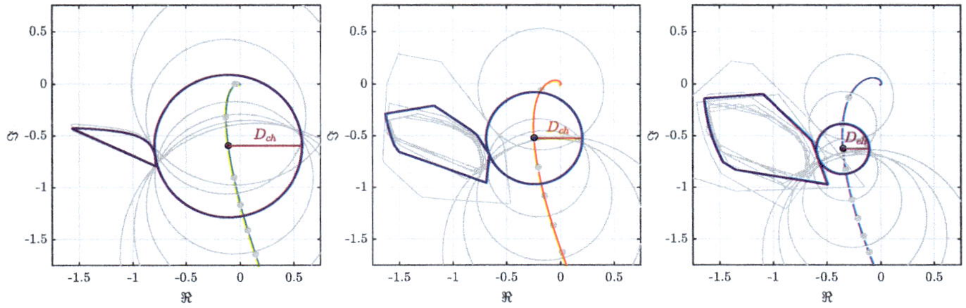

(a) T_s = 200 *ms*, D_{ch} = 0.69 (b) T_s = 14 *s*, D_{ch} = 0.45 (c) T_s = 28 *s*, D_{ch} = 0.25

Fig. 8.15 Sampled DF analysis for SHR controller

According to the rule of thumb to choose the control period, the T_s chosen for the previous section is excessive and, since large robustness margin is available, greater control periods can be chosen. For the sake of the experiment it has been chosen two additional control periods of $T_s = 14$ s, which correspond to the upper limit of the rule of thumb, and $T_s = 28$ s, which falls outside the typical range.

As in the case with $T_s = 200$ ms the Sampled DF approach has been applied to the new cases. Each of the normalized open-loop transfer functions has been represented with the convex hull encircling the sampled DF traces for a range of values of r in Fig. 8.15b, c. As it can be seen in these figures in both cases limit cycle oscillations are avoided since $D_{ch} > 0$. For the case $T_s = 14$ s a value $D_{ch} = 0.45$ is obtained for $r = 13$ and for the case $T_s = 28$ s a value $D_{ch} = 0.25$ is obtained for $r = 8$. Nevertheless, it can be observed how the robustness measure D_{ch} lowers when T_s increases, thus, increasing the control period further could result in the apparition of limit cycle oscillations.

The SHR controller has been tested on the system changing the control periods. The responses of the system against a step change in the reference input and their corresponding control action can be found in Fig. 8.16. And the responses against a disturbance input in Fig. 8.17 with their corresponding control action. As expected the controller avoids limit cycle oscillations with all the considered control periods.

This kind of experiments reducing the control frequency can decrease significantly the number of control action changes, however, the magnitude of these changes increases. This reduction of control action changes may be beneficial for the actuator.

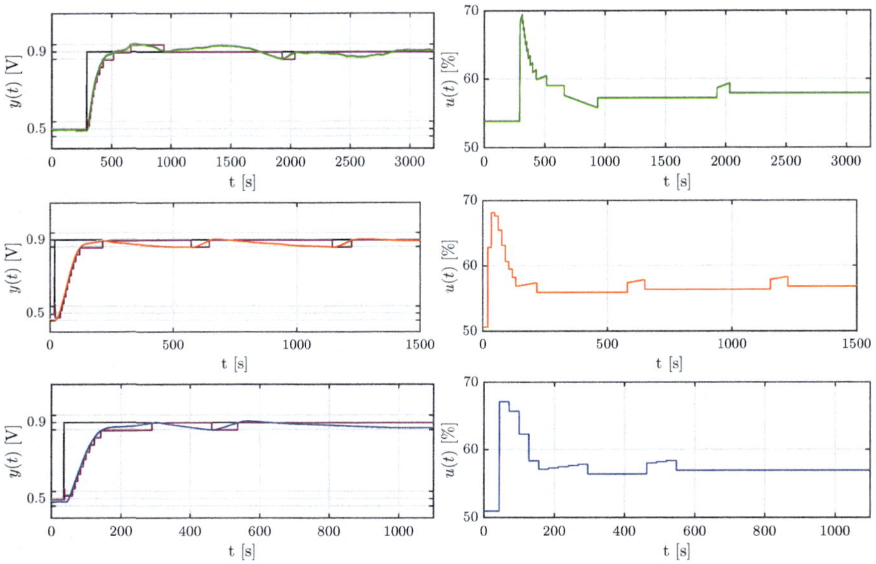

Fig. 8.16 Step response of the studied system with different control periods and respective control action. $T_s = 200\,\text{ms}$ in dark green, $T_s = 14\,\text{s}$ in red and $T_s = 28\,\text{s}$ in blue

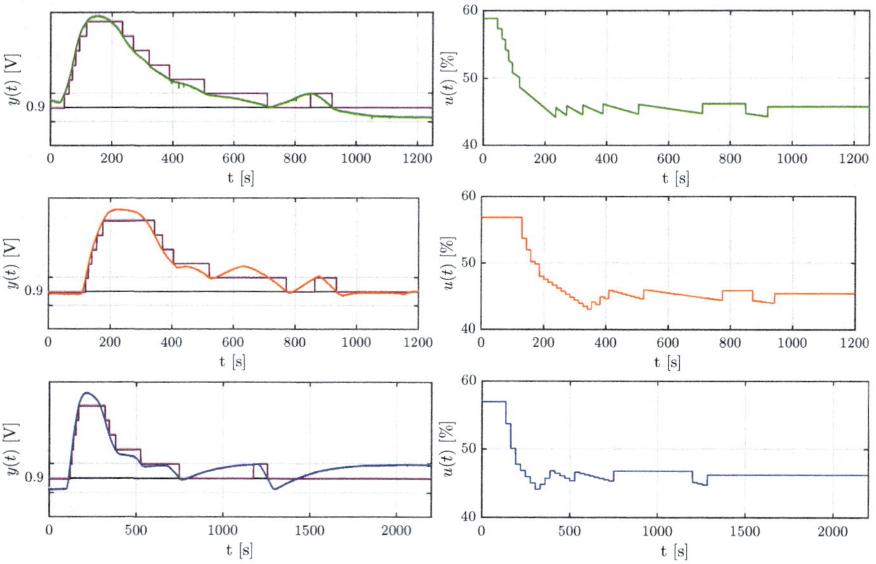

Fig. 8.17 Response against disturbance of the studied system with different control periods and respective control action. $T_s = 200\,\text{ms}$ in dark green, $T_s = 14\,\text{s}$ in red and $T_s = 28\,\text{s}$ in blue

8.5.5 Summary

In this section, a summary of the control experiences is presented. In Table 8.2 a summary of the controllers obtained with different tuning methods with the robustness that they provide is shown.

The robustness parameters presented include classical phase and gain margins (Φ_{cp}, γ_{cg}), specific gain and phase margins to the non-linearity under study ($\Phi_{h/\delta}$, $\gamma_{h/\delta}$) and the Tsypkin's robustness margin (M_T) to the non-linearity in the loop.

The results shown in this table indicate that there exists certain correlation between classical robustness margins and the specific margins, since the higher the classical margins are, the more the specific margins increase. Nevertheless, despite this correlation, classical robustness margins cannot be taken as a real indicative for the behavior of the system when the type of non-linearities under study are implemented in the control loop, since they do not determine the robustness against the limit cycle oscillations that these non-linearities induce.

As it is known, the DF technique is applicable as long as the input to the non-linear part of the loop reassembles a sinusoid and the linear part of the system filters sufficiently the output signal to consider relevant only the first harmonic, i.e. a sinusoid, that will be reintroduced at the input of the non-linear part. Therefore, the robustness margins obtained from this method will remain valid as long as this filtering hypothesis is fulfilled.

In the experimental results section, it has been showed how, despite the apparent robustness against limit cycle oscillations that the DF assured, some sustained oscillations appeared. As it has been shown lately, Tsypkin's method predicted with more accuracy all the cases. The unique conflictive case is the AMIGO controller with a SSOD sampling strategy in which, even if the robustness measure M_T is greater than

Table 8.2 Summary of robustness margins provided by the controllers under study

	SLR		SHR		AMIGO	Ziegler	Tsypkin
	SSOD	RQH	SSOD	RQH		Nichols	
K_p	157.37	157.37	37.82	93.76	45.96	142.05	52.51
T_i	44.76	44.76	221.68	107.76	80.36	36.11	90.76
Φ_{cp} [°]	45	45	88	65.7	67.6	42.5	69.6
γ_{cg} $[dB]$	15	15	28.5	20.4	26.4	15.6	25.3
$\Phi_{h/\delta}$ [°] (SSOD)	1.7	–	45	–	23	−1.3	25.2
$\gamma_{h/\delta}$ $[dB]$ (SSOD)	0.92	–	17.4	–	14.3	−1.2	13.4
$\Phi_{h/\delta}$ [°] (RQH)	–	24.1	–	45	44.8	21	47.2
$\gamma_{h/\delta}$ $[dB]$ (RQH)	–	9.4	–	15	21	9.7	20
M_T (SSOD)	0	–	0.71	–	0.15	0	0.5
M_T (RQH)	–	0	–	0.53	0.63	0	0.62

Table 8.3 Summary of detected oscillation and validity of the DF approach

	SLR		SHR		AMIGO	Ziegler Nichols	Tsypkin
	SSOD	RQH	SSOD	RQH			
SSOD oscillation	Yes	–	No	–	Yes	Yes	No
SSOD DF accurate	Yes	–	Yes	–	No	Yes	Yes
RQH oscillation	–	Yes	–	No	No	Yes	No
RQH DF accurate	–	No	–	Yes	Yes	No	Yes

0, it is also true that the obtained value is found for a low frequency, and therefore, a small variation in the obtained model could alter significantly its value.

In Table 8.3, a summary of the experimental results and their relation with the analysis can be found.

The table gathers whether an oscillation has been found or not for different sampling strategies and the controllers obtained with different methods. Then, an entry is included reflecting if the initial DF method used to analyze them has been accurate. It is considered as accurate the cases where the trace of the inverse negative of the DF for $m = 1$ reassembles the critical Tsypkin Branch.

With regard to the control period refreshment experiments, the sampled DF has proven its validity since for the studied cases no limit cycles have appeared as it was expected since $D_{ch} > 0$ was obtained for all the cases. In addition, for the case that can be considered as continuous, the obtained D_{ch} is very similar to the Tsypkin Margin, and both are measured from a similar frequency. This fact shows the convergence of the Sampled DF to the continuous case when the control frequency is high enough to be neglected with regard to the system dynamics.

Quick summary:

- ☐ Implementation of IEC 61499 applications for continuous processes has been carried out.
- ☐ Development of a networked EBC application leveraging the standard's features.
- ☐ Ad-hoc FB's for implementing RQH, SSOD and additional functionalities have been developed.
- ☐ The setup of a tank volume control experiment forming a networked control system is presented.
- ☐ Experimental application of theoretical content covered in previous chapters is demonstrated.
- ☐ Distributed EBC is experimentally validated as a successful control strategy, supported by a solid theoretical background.

References

1. Dormido S, Sánchez J, Kofman E (2008) Muestreo, control y comunicación basados en eventos. Revista Iberoamericana de Automática e Informática Industrial RIAI 5(1):5–26
2. Årzén K-E (1999) A simple event-based PID controller. In: Proceedings of the 14th IFAC world congress, vol 18, pp 423–428
3. Durand S, Marchand N (2009) An event-based PID controller with low computational cost. In: 8th international conference on sampling theory and applications (SampTA'09), pages Special–session
4. Durand S, Marchand N (2009) Further results on event-based PID controller. In: Control conference (ECC), 2009 European, pp 1979–1984. IEEE
5. Vasyutynskyy V, Kabitzsch K (2009) Time constraints in PID controls with send-on-delta. IFAC Proc Vol 42(3):48–55
6. Vasyutynskyy V, Kabitzsch K (2010) A comparative study of PID control algorithms adapted to send-on-delta sampling. In: 2010 IEEE international symposium on industrial electronics (ISIE), pp 3373–3379. IEEE
7. Beschi M, Dormido S, Sánchez J, Visioli A (2012) Characterization of symmetric send-on-delta PI controllers. J Process Control 22(10):1930–1945
8. Miguel-Escrig O, Romero-Pérez J-A (2020) Regular quantisation with hysteresis: a new sampling strategy for event-based PID control systems. IET Control Theory Appl
9. Beschi M, Dormido S, Sánchez J, A Visioli (2014) Tuning of symmetric send-on-delta proportional-integral controllers. IET Control Theory Appl 8:248–259(11)
10. Romero-Pérez J-A, Sanchis-Llopis R (2016) A new method for tuning PI controllers with symmetric send-on-delta sampling strategy. ISA Trans 64:161–173
11. Romero-Pérez J-A, Sanchis-Llopis R (2016) Analysis of a simple rule for tuning SSOD based PIDs. In: 2016 second international conference on event-based control, communication, and signal processing (EBCCSP), pp 1–8. IEEE
12. Miguel-Escrig O, Romero-Pérez J-A, Sanchis-Llopis R (2020) Tuning PID controllers with symmetric send-on-delta sampling strategy. J Franklin Inst 357(2):832–862
13. Romero-Pérez J-A, Sanchis-Llopis R (2017) Tuning and robustness analysis of event-based PID controllers under different event generation strategies. Int J Control 1–38
14. Tsividis Y (2010) Event-driven data acquisition and digital signal processing—A tutorial. IEEE Trans Circuits Syst II: Express Briefs 57(8):577–581
15. Sayiner N, Sorensen HV, Viswanathan TR (1996) A level-crossing sampling scheme for A/D conversion. IEEE Trans Circuits Syst II: Analog Digital Signal Process 43(4):335–339
16. John K-H, Tiegelkamp M (2010) IEC 61131-3: programming industrial automation systems: concepts and programming languages, requirements for programming systems, decision-making aids. Springer
17. Zoitl A, Lewis R (2014) Modelling control systems using IEC 61499, vol 95. IET
18. Eclipse 4diac - the open source environment for distributed industrial automation and control systems, 2007–2021
19. Ziegler JG, Nichols NB (1942) Optimum settings for automatic controllers. Trans ASME 64(11)
20. Åström KJ, Hägglund T (2004) Revisiting the Ziegler-Nichols step response method for PID control. J Process Control 14(6):635–650

Appendix A
Temporal Response Induced by a Multileveled Sampler

Consider the sampled error signal $\bar{e}(t)$ follows a ladder-type form as described in Sect. 4.3. The expression of $\bar{e}(t)$ is obtained through Fourier series expansion:

$$\bar{e}(t) = \frac{a_0}{2} + \sum_{n=1}^{\infty} (a_n \cos(\omega_n t) + b_n \sin(\omega_n t)). \tag{A.1}$$

As the error signal is supposed unbiased and symmetric the coefficient a_0 is equal to 0. The remaining coefficients:

$$a_n = \frac{2}{T_o} \int_{\rho_1 \frac{T_o}{2}}^{\rho_2 \frac{T_o}{2}} \delta \cos(\omega_n t) dt + \frac{2}{T_o} \int_{\rho_2 \frac{T_o}{2}}^{\rho_3 \frac{T_o}{2}} 2\delta \cos(\omega_n t) dt + \cdots + \frac{2}{T_o} \int_{\rho_m \frac{T_o}{2}}^{\rho_{m+1} \frac{T_o}{2}} m\delta \cos(\omega_n t) dt +$$
$$+ \frac{2}{T_o} \int_{\rho_{m+1} \frac{T_o}{2}}^{\rho_{m+2} \frac{T_o}{2}} (m-1)\delta \cos(\omega_n t) dt + \cdots + \frac{2}{T_0} \int_{\rho_{2m-1} \frac{T_0}{2}}^{\rho_{2m} \frac{T_o}{2}} \delta \cos(\omega_n t) dt +$$
$$+ \frac{2}{T_o} \int_{\frac{T_o}{2} + \rho_1 \frac{T_o}{2}}^{\frac{T_o}{2} + \rho_2 \frac{T_o}{2}} -\delta \cos(\omega_n t) dt + \cdots + \frac{2}{T_o} \int_{\frac{T_o}{2} + \rho_m \frac{T_o}{2}}^{\frac{T_o}{2} + \rho_{m+1} \frac{T_o}{2}} -m\delta \cos(\omega_n t) dt +$$
$$+ \frac{2}{T_o} \int_{\frac{T_o}{2} + \rho_{m+1} \frac{T_o}{2}}^{\frac{T_o}{2} + \rho_{m+2} \frac{T_o}{2}} -(m-1)\delta \cos(\omega_n t) dt + \cdots + \frac{2}{T_o} \int_{\frac{T_o}{2} + \rho_{2m-1} \frac{T_o}{2}}^{\frac{T_o}{2} + \rho_{2m} \frac{T_o}{2}} -\delta \cos(\omega_n t) dt$$

Solving and arranging the equation:

O. Miguel-Escrig, *Event-Based PID Controllers with Fixed Threshold Sampling Strategies*, Springer Theses, https://doi.org/10.1007/978-3-031-62505-3

$$a_n = \begin{cases} 0 & \text{if n is even} \\ \frac{2\delta}{n\pi}\left(\sum_{i=1}^{m} i(\sin(n\pi\rho_{i+1}) - \sin(n\pi\rho_i)) + \sum_{i=m+1}^{2m-1}(2m-i)(\sin(n\pi\rho_{i+1}) - \sin(n\pi\rho_i))\right) & \text{if } n \text{ is odd} \end{cases}$$

Similarly for b_n:

$$b_n = \begin{cases} 0 & \text{if } n \text{ is even} \\ \frac{2\delta}{n\pi}\left(\sum_{i=1}^{m} i(\cos(n\pi\rho_i) - \cos(n\pi\rho_{i+1})) + \sum_{i=m+1}^{2m-1}(2m-i)(\cos(n\pi\rho_i) - \cos(n\pi\rho_{i+1}))\right) & \text{if } n \text{ is odd} \end{cases}$$

Substituting a_0, a_n and b_n in Eq. (A.1), $\bar{e}(t)$ is obtained. Considering that $e = -L\bar{e}$, and knowing that L is the open-loop transfer function of the system ($L = G_{ol}(j\omega)$), the general expression for $e(t)$ results in:

$$\begin{aligned} e\left(\rho_p \frac{T_o}{2}\right) = & -\frac{4\delta}{\pi}\sum_{n_{odd}}^{\infty}\frac{1}{n}\left(\Re\{G_{ol}(jn\omega_o)\}\left(\sum_{i=1}^{m} i\sin\left(\frac{n\pi}{2}(\rho_{i+1} - \rho_i)\right)\cos\left(n\pi\left(\frac{\rho_{i+1} + \rho_i}{2} - \rho_p\right)\right)\right.\right. \\ & \left.\left. + \sum_{i=m+1}^{2m-1}(2m-i)\sin\left(\frac{n\pi}{2}(\rho_{i+1} - \rho_i)\right)\cos\left(n\pi\left(\frac{\rho_{i+1} + \rho_i}{2} - \rho_p\right)\right)\right)\right) \\ & -\frac{4\delta}{\pi}\sum_{n_{odd}}^{\infty}\frac{1}{n}\left(\Im\{G_{ol}(jn\omega_o)\}\left(\sum_{i=1}^{m} i\sin\left(\frac{n\pi}{2}(\rho_{i+1} - \rho_i)\right)\sin\left(n\pi\left(\frac{\rho_{i+1} + \rho_i}{2} - \rho_p\right)\right)\right.\right. \\ & \left.\left. + \sum_{i=m+1}^{2m-1}(2m-i)\sin\left(\frac{n\pi}{2}(\rho_{i+1} - \rho_i)\right)\sin\left(n\pi\left(\frac{\rho_{i+1} + \rho_i}{2} - \rho_p\right)\right)\right)\right) \end{aligned}$$

Appendix B
DF Related Calculations

B.1 RQH Calculation

The output of the proposed RQH sampler is given by:

$$\bar{e}(t) = \begin{cases} (i+1)\delta \text{ if } e(t) \geq \left(i + \frac{1}{2} + \frac{h}{2\delta}\right)\delta \text{ and } \bar{e}(t^-) = i\delta, \; i \in \mathbb{Z} \\ (i-1)\delta \text{ if } e(t) \leq \left(i - \frac{1}{2} - \frac{h}{2\delta}\right)\delta \text{ and } \bar{e}(t^-) = i\delta \\ i\delta \qquad \text{ if } e(t) \in \left[\left(i - \frac{1}{2} - \frac{h}{2\delta}\right)\delta, \left(i + \frac{1}{2} + \frac{h}{2\delta}\right)\delta\right] \text{ and } \bar{e}(t^-) = i\delta \end{cases}$$

For a sinusoidal input $e(\phi) = A\sin(\phi)$, the output of the sampler can be expressed as:

$$\begin{aligned} \bar{e}(\phi) &= \delta \sum_{k=1}^{i} \operatorname{sgn}\left(\left.\frac{de(\phi)}{d\phi}\right|_{\phi_k}\right) \quad \forall\phi;\, \phi_i < \phi < \phi_{i+1} \\ &= \delta \sum_{k=1}^{i} \operatorname{sgn}\left(\cos(\phi_k)\right). \end{aligned} \tag{B.1}$$

The EG-ZOH sampler is an odd non-linearity, and the past values of the input determine the current value of the output in the multiple-valued regions. The Describing Function for this kind of non-linearity is calculated as:

O. Miguel-Escrig, *Event-Based PID Controllers with Fixed Threshold Sampling Strategies*, Springer Theses, https://doi.org/10.1007/978-3-031-62505-3

$$\begin{aligned}\mathcal{N}(A,h) &= \frac{2j}{\pi A}\int_0^{\pi}\bar{e}(\phi)e^{-j\phi}d\phi \\ &= \frac{2j}{\pi A}\left(\int_{\phi_1}^{\phi_2}\delta\,\mathrm{sgn}(\cos\phi_1)e^{-j\phi}d\phi + \int_{\phi_2}^{\phi_3}\delta(\mathrm{sgn}(\cos\phi_1)+\right. \\ &\left.\mathrm{sgn}(\cos\phi_2))e^{-j\phi}d\phi + \cdots + \int_{\phi_n}^{\pi}\delta\sum_{k=1}^{n}(\mathrm{sgn}(\cos\phi_k))e^{-j\phi}d\phi\right)\end{aligned} \tag{B.2}$$

which can be rewritten as:

$$\begin{aligned}\mathcal{N}(A,h) &= \frac{2\delta j}{\pi A}\left(\int_{\phi_1}^{\pi}\mathrm{sgn}(\cos\phi_1)e^{-j\phi}d\phi + \int_{\phi_2}^{\pi}\mathrm{sgn}(\cos\phi_2)e^{-j\phi}d\phi + \right. \\ &\left.\cdots + \int_{\phi_n}^{\pi}\mathrm{sgn}(\cos\phi_n)e^{-j\phi}d\phi\right) \\ &= \frac{2\delta j}{\pi A}\sum_{k=1}^{n}\left(\mathrm{sgn}(\cos\phi_k)\int_{\phi_k}^{\pi}e^{-j\phi}d\phi\right).\end{aligned} \tag{B.3}$$

Taking into account that:

$$\int_{\phi_k}^{\pi}e^{-j\phi}d\phi = -\sin\phi_k - j(1+\cos\phi_k), \tag{B.4}$$

equation (B.3) results in:

$$\mathcal{N}(A,h) = \frac{2\delta}{\pi A}\left(\sum_{k=1}^{n}(1+\cos\phi_k)\,\mathrm{sgn}(\cos\phi_k) - j\sum_{k=1}^{n}\sin\phi_k\,\mathrm{sgn}(\cos\phi_k)\right), \tag{B.5}$$

which can be transformed to:

$$\mathcal{N}(A,h) = \frac{2\delta}{\pi A}\left(\sum_{k=1}^{m}\cos\phi_k - \sum_{k=m+1}^{2m}\cos\phi_k - j\left(\sum_{k=1}^{m}\sin\phi_k - \sum_{k=m+1}^{2m}\sin\phi_k\right)\right). \tag{B.6}$$

As the expressions of the $\sin\phi_k$ and $\cos\phi_k$ are known to be:

$$\sin\phi_k = \begin{cases}\dfrac{\delta}{A}\left(k-\dfrac{1}{2}+\dfrac{h}{2\delta}\right) & \text{if } k = 1,2,\ldots,m \\ \dfrac{\delta}{A}\left(2m-k+\dfrac{1}{2}-\dfrac{h}{2\delta}\right) & \text{if } k = m+1, m+2, \ldots, 2m\end{cases} \tag{B.7}$$

$$\cos\phi_k = \begin{cases} +\sqrt{1-\left(\frac{\delta}{A}\left(k-\frac{1}{2}+\frac{h}{2\delta}\right)\right)^2} & \text{if } k=1,2,\ldots,m \\ -\sqrt{1-\left(\frac{\delta}{A}\left(2m-k+\frac{1}{2}-\frac{h}{2\delta}\right)\right)^2} & \text{if } k=m+1,\ m+2,\ldots,2m \end{cases} \tag{B.8}$$

can be introduced in Eq. (B.6) which results in:

$$\mathcal{N}(A,h) = \frac{2\delta}{A\pi}\left[\sum_{k=1}^{m}\sqrt{1-\left(\frac{\delta}{A}\left(k+\frac{h}{2\delta}-\frac{1}{2}\right)\right)^2} + \sum_{k=m+1}^{2m}\sqrt{1-\left(\frac{\delta}{A}\left(2m-k-\frac{h}{2\delta}+\frac{1}{2}\right)\right)^2}\right] - j\frac{2hm\delta}{A^2\pi} \tag{B.9}$$

B.2 DIDF Calculation

The input-output relationship of a FTS, described as in Fig. 3.2, is given by:

$$\bar{x}(t) = \begin{cases} \Delta\epsilon - \frac{\epsilon}{2} + \epsilon(i+1) & \text{if } x(t) \geq \Delta\delta + \frac{h}{2} + \delta i \text{ and } \bar{x}(t^-) = \Delta\epsilon - \frac{\epsilon}{2} + \epsilon i,\quad i \in \mathbb{Z} \\ \Delta\epsilon - \frac{\epsilon}{2} + \epsilon(i-1) & \text{if } x(t) \leq \Delta\delta - \frac{h}{2} + \delta(i-1) \text{ and } \bar{x}(t^-) = \Delta\epsilon - \frac{\epsilon}{2} + \epsilon i \\ \Delta\epsilon - \frac{\epsilon}{2} + \epsilon i & \text{if } x(t) \in \left[\Delta\delta - \frac{h}{2} + \delta(i-1), \Delta\delta + \frac{h}{2} + \delta i\right] \text{ and } \bar{x}(t^-) = \Delta\epsilon - \frac{\epsilon}{2} + \epsilon i \end{cases}$$

Let the input to the FTS be described by $x(\phi) = A\sin(\phi) + B$. Then, the output can be expressed as:

$$\begin{aligned} \bar{x}(\phi) &= \Delta\epsilon - \frac{\epsilon}{2} + \epsilon i_0 + \epsilon\sum_{k=1}^{i}\operatorname{sgn}\left(\left.\frac{dx(\phi)}{d\phi}\right|_{\phi_k}\right) \quad \forall\phi;\ \phi_i < \phi < \phi_{i+1} \\ &= \Delta\epsilon - \frac{\epsilon}{2} + \epsilon i_0 + \epsilon\sum_{k=1}^{i}\operatorname{sgn}(\cos(\phi_k)), \end{aligned}$$

where i_0 is the level around which the oscillation is centered. Under this assumption for the form of $x(\phi)$ and because of a possible output bias, the Dual Input Describing Function (DIDF) must be used. This variant of the Describing Function is composed of one part related to the sustainability of the bias, and another one related to the dynamic part, which can be respectively calculated with the following expressions:

$$\mathcal{N}_B = \frac{1}{2\pi B}\int_0^{2\pi}\bar{x}(\phi)d\phi$$

and

$$\mathcal{N}_A = \frac{j}{\pi A}\int_0^{2\pi} \bar{x}(\phi)e^{-j\phi}d\phi.$$

Developing, for example, for $\mathcal{N}_A$:

$$\mathcal{N}_A = \frac{j}{\pi A}\Bigg[\int_0^{\phi_1}\left(\Delta\epsilon - \frac{\epsilon}{2} + \epsilon i_0\right)e^{-j\phi}d\phi + \int_{\phi_1}^{\phi_2}\left(\Delta\epsilon - \frac{\epsilon}{2} + \epsilon i_0 + \epsilon\,\mathrm{sgn}(\cos(\phi_1))\right)e^{-j\phi}d\phi + \cdots$$
$$+ \int_{\phi_{2m_{sup}+2m_{inf}-1}}^{\phi_{2m_{sup}+2m_{inf}}}\left(\Delta\epsilon - \frac{\epsilon}{2} + \epsilon i_0 + \epsilon\sum_{k=1}^{2m_{sup}+2m_{inf}-1}\mathrm{sgn}(\cos(\phi_k))\right)e^{-j\phi}d\phi + ,$$
$$+ \int_{\phi_{2m_{sup}+2m_{inf}}}^{2\pi}\left(\Delta\epsilon - \frac{\epsilon}{2} + \epsilon i_0 + \epsilon\sum_{k=1}^{2m_{sup}+2m_{inf}}\mathrm{sgn}(\cos(\phi_k))\right)e^{-j\phi}d\phi\Bigg]$$

where m_{sup} and m_{inf} are the number of levels crossed in upwards and downwards direction. Arranging terms:

$$\mathcal{N}_A = \frac{j}{\pi A}\left[\left(\Delta\epsilon - \frac{\epsilon}{2} + \epsilon i_0\right)\int_0^{2\pi} e^{-j\phi}d\phi + \epsilon\sum_{k=1}^{2m_{sup}+2m_{inf}}\int_{\phi_k}^{2\pi}\mathrm{sgn}(\cos(\phi_k))e^{-j\phi}d\phi\right].$$

The first integral equals 0, and for the second, it is known that:

$$\mathrm{sgn}(\cos(\phi_i)) = \begin{cases} +1 & 1 \le i \le m_{sup} \\ -1 & m_{sup} < i \le 2m_{sup} + m_{inf} \\ +1 & 2m_{sup} + m_{inf} < i \le 2m_{sup} + 2m_{inf} \end{cases}$$

After some straightforward algebra calculations and arrangements the following expression is found:

$$\mathcal{N}_A = \frac{\epsilon}{\pi A}\sum_{k=-m_{inf}}^{m_{sup}-1}\left[\sqrt{1-\left(\frac{1}{A}\left(\Delta\delta + \frac{h}{2} - B + \delta(k+i_0)\right)\right)^2} + \sqrt{1-\left(\frac{1}{A}\left(\Delta\delta - \frac{h}{2} - B + \delta(k+i_0)\right)\right)^2}\right]$$
$$- j\frac{\epsilon h}{\pi A^2}[m_{sup} + m_{inf}].$$

The same procedure is followed to obtain the expression of $\mathcal{N}_B$:

$$\mathcal{N}_B = \frac{2\Delta\epsilon + \epsilon(m_{sup} - m_{inf} + 2i_0 - 1)}{2B}$$
$$- \frac{\epsilon}{2\pi B}\sum_{k=-m_{inf}}^{m_{sup}-1}\left[\arcsin\left(\frac{1}{A}\left(\Delta\delta + \frac{h}{2} - B + \delta(k+i_0)\right)\right) + \arcsin\left(\frac{1}{A}\left(\Delta\delta - \frac{h}{2} - B + \delta(k+i_0)\right)\right)\right].$$

The expressions for m_{sup} and m_{inf} can be calculated with:

$$m_{sup} = \left\lfloor \frac{A + B - \Delta\delta - h/2 - \delta i_0}{\delta} \right\rfloor + 1, \quad m_{inf} = -\left\lfloor \frac{-A + B - \Delta\delta + h/2 - \delta i_0}{\delta} \right\rfloor - 1$$

and the initial level i_0 with:

$$i_0 = \begin{cases} \left\lfloor \dfrac{B - \Delta\delta - h/2}{\delta} \right\rfloor + 1 & \text{if} \quad B \geq 0 \\ \left\lfloor \dfrac{B - \Delta\delta + h/2}{\delta} \right\rfloor & \text{if} \quad B < 0 \end{cases}.$$

B.3 DIDF Maintenance of the Bias

Depending on the process and controller in the loop there exist different situations which define the necessary conditions to obtain a sustained bias. The disturbance and reference signals have been considered to be step-like signals of magnitude $p(\infty)$ and $r(\infty)$ respectively, resulting in the oscillation conditions to be a function of the process and controller involved. Then, the different combinations between these elements and their respective oscillation condition are presented in Table B.1.

Table B.1 Expanded requirements for any combination of process and controller for the maintenance of the bias under step-like changes in reference and disturbance inputs. $L_1(s)$ and $L_2(s)$ have neither poles nor zeros at the origin and it is considered that $n, m \geq 1$

Form of $C(s)$ and $G(s)$			Conditions
$C(s) = L_1(s)s^n$	$G(s) = L_2(s)s^m$		$B = r(\infty)$
	$G(s) = L_2(s)$		$B = r(\infty) - p(\infty)L_2(0)$
	$G(s) = L_2(s)/s^m$	$n > m$	$p(\infty) = 0$ & $B = r(\infty)$
		$n = m$	$p(\infty) = 0$ & $B(1 + \mathcal{N}_B L_1(0)L_2(0)) = r(\infty)$
		$n < m$	$p(\infty) = 0$ & $B = 0$
$C(s) = L_1(s)$	$G(s) = L_2(s)s^m$		$B = r(\infty)$
	$G(s) = L_2(s)$		$B(1 + \mathcal{N}_B L_1(0)L_2(0)) = r(\infty) - p(\infty)L_2(0)$
	$G(s) = L_2(s)/s^m$		$B\mathcal{N}_B = -p(\infty)/L_1(0)$
$C(s) = L_1(s)/s^n$	$G(s) = L_2(s)/s^m$		$B = 0$
	$G(s) = L_2(s)$		$B = 0$
	$G(s) = L_2(s)s^m$	$n < m$	$B = r(\infty)$
		$n = m$	$B(1 + \mathcal{N}_B L_1(0)L_2(0)) = r(\infty)$
		$n > m$	$B = 0$

B.4 Sampled DF Calculation

The sampled describing function which relates the input and output of the non-linear element in the system can be computed in the following way:

$$\mathcal{N}(\delta, A, T_s) = \frac{\text{Phasor representation of fundamental component of } \bar{e}^*}{\text{Phasor representation of } e}.$$

Firstly, the phasor representation of e can be easily obtained since for the DF calculations:

$$e(t) = A\sin(\omega_o t) = A\cos\left(\omega_o t + \frac{3\pi}{2}\right) = \Re\left\{Ae^{j\left(\omega_o t + \frac{3\pi}{2}\right)}\right\}.$$

For the phasor representation of the fundamental component of $\bar{e}^*$, an harmonic analysis using Fourier series has been done. Expressing $\bar{e}^*(t)$ as:

$$\bar{e}^*(t) = \bar{e}(t)\cdot\delta_{PT}(t),$$

where the pulse train δ_{PT} is defined as:

$$\delta_{PT}(t) = \sum_{k=-\infty}^{\infty}\delta_D(t - \tau - kT_s),$$

where δ_D is the Dirac delta function, T_s is the sampling period and τ the time lag between the initial zero-crossing of $e(t)$ and the first sample (which is bounded between 0 and T_s).

To obtain the fundamental component of $\bar{e}^*(t)$, firstly, the Fourier series representing $\bar{e}(t)$ is obtained:

$$\hat{\bar{e}}(t) = -\frac{\delta}{j\pi}\sum_{n=-\infty}^{\infty}\frac{1}{n}\left\{\sum_{i=1}^{m} i\left[e^{-jn\omega_o t_{i+1}} - e^{-jn\omega_o t_i}\right] + \sum_{i=m+1}^{2m-1}(2m-i)\left[e^{-jn\omega_o t_{i+1}} - e^{-jn\omega_o t_i}\right]\right\}e^{jn\omega_o t},$$

where t_n are the times where level switches are produced. And secondly, the Fourier series representation of δ_{PT} is obtained:

$$\hat{\delta}_{PT}(t) = \frac{1}{T_s}\sum_{k=-\infty}^{\infty} e^{jk\omega_s(t-\tau)}.$$

Multiplying both:

$$\hat{\bar{e}}^*(t) = -\frac{\delta}{j\pi T_s}\sum_{k=-\infty}^{\infty}\sum_{n=-\infty}^{\infty}\frac{1}{n}\left\{\sum_{i=1}^{m} i\left[e^{-jn\omega_o t_{i+1}} - e^{-jn\omega_o t_i}\right] + \sum_{i=m+1}^{2m-1}(2m-i)\left[e^{-jn\omega_o t_{i+1}} - e^{-jn\omega_o t_i}\right]\right\} e^{jn\omega_o t} e^{jk\omega_s (t-\tau)}.$$

To obtain the fundamental harmonic from this expression, attention has to be paid to the exponents that imply the variable t, which can be grouped in a single expression:

$$j(n\omega_o + k\omega_s)t - jk\omega_s\tau.$$

Then, the part that multiplies t has to be the fundamental frequency, i.e. either $+\omega_o$ or $-\omega_o$. Taking r as the ratio between the oscillation and sampling period ($r = T_o/T_s$), the relation between the harmonics of the sampling (k) and of the signal $\bar{e}(t)$ (n) can be obtained to calculate the fundamental frequency of $\bar{e}^*(t)$:

$$\begin{aligned} n\omega_o + k\omega_s &= \omega_o & \quad n\omega_o + k\omega_s &= -\omega_o \\ n + k\frac{\omega_s}{\omega_o} &= 1 & \quad n + k\frac{\omega_s}{\omega_o} &= -1 \\ n &= 1 - kr & \quad n &= -1 - kr \end{aligned}$$

Thus, the relation of harmonics to consider are both $n = 1 - kr$ and $n = -1 - kr$. The expression of the fundamental harmonic of $\bar{e}^*(t)$ is:

$$\begin{aligned}\hat{\bar{e}}^*(t) = -\frac{\delta}{j\pi T_s}\sum_{k=-\infty}^{\infty}\Bigg\{&\frac{1}{1-rk}\left\{\sum_{i=1}^{m} i\left[e^{-j(1-rk)\omega_o t_{i+1}} - e^{-j(1-rk)\omega_o t_i}\right] + \sum_{i=m+1}^{2m-1}(2m-i)\left[e^{-j(1-rk)\omega_o t_{i+1}} - e^{-j(1-rk)\omega_o t_i}\right]\right\} e^{j\omega_o t} + \\ &\frac{1}{-1-rk}\left\{\sum_{i=1}^{m} i\left[e^{-j(-1-rk)\omega_o t_{i+1}} - e^{-j(-1-rk)\omega_o t_i}\right] + \sum_{i=m+1}^{2m-1}(2m-i)\left[e^{-j(-1-rk)\omega_o t_{i+1}} - e^{-j(-1-rk)\omega_o t_i}\right]\right\} e^{-j\omega_o t}\Bigg\} e^{-jk\omega_s\tau}.\end{aligned}$$

Taking into account that:

$$e^{-j(1-rk)\omega_o t_i} = \left(e^{j\omega_o t_i}\right)^{rk-1} = \left(\cos(\omega t_i) + j\sin(\omega_o t_i)\right)^{rk-1}$$

and:

$$\sin(\omega_o t_i) = \begin{cases} \dfrac{i\delta}{A} & \text{if} \quad i = 1, 2, \ldots, m \\ (2m - i)\,\dfrac{\delta}{A} & \text{if} \quad i = m + 1, m + 2, \ldots, 2m \end{cases}$$

$$\cos(\omega_o t_i) = \begin{cases} +\sqrt{1 - \left(\dfrac{i\delta}{A}\right)^2} & \text{if} \quad i = 1, 2, \ldots, m \\ -\sqrt{1 - \left((2m - i)\,\dfrac{\delta}{A}\right)^2} & \text{if} \quad i = m + 1, m + 2, \ldots, 2m \end{cases}$$

the expression of $\hat{e}^*(t)$ can be simplified and its phasor representation obtained after some straightforward algebra calculus. Then, the ratio between the phasor of $\hat{e}^*(t)$ and the phasor of $e(t)$ resulting in the Sampled DF:

$$\mathcal{N}(\delta/A, r, \tau, T_s) = \frac{2\delta}{\pi T_s A} \sum_{k=-\infty}^{\infty} \frac{e^{-jk2\pi\tau/T_s}}{rk - 1} \left[\sum_{i=1}^{m-1} \left\{ \left(-\sqrt{1 - \left(\frac{i\delta}{A}\right)^2} + j\frac{i\delta}{A} \right)^{rk-1} \right.\right.$$
$$\left.\left. - \left(\sqrt{1 - \left(\frac{i\delta}{A}\right)^2} + j\frac{i\delta}{A} \right)^{rk-1} \right\} - \left(\sqrt{1 - \left(\frac{m\delta}{A}\right)^2} + j\frac{m\delta}{A} \right)^{rk-1} + (-1)^{rk-1} \right].$$

Appendix C
Tuning Methods

C.1 Tuning Table for Optimum SSOD

See Table C.1.

C.2 Flowchart for RQH Tuning Procedure

The proposed tuning procedure can be deployed by following the flowchart in Fig. C.1.

O. Miguel-Escrig, *Event-Based PID Controllers with Fixed Threshold Sampling Strategies*, Springer Theses, https://doi.org/10.1007/978-3-031-62505-3

Table C.1 Tuning table for $M_{T_r} > 0.2$. (∗): Take the parameters for the precedent case of $\overline{C}(\infty)$

L/τ	$\overline{C}(\infty) < 1$			$\overline{C}(\infty) < 2$			$\overline{C}(\infty) < 5$			$\overline{C}(\infty) < 10$			$\forall\overline{C}(\infty)$		
	$\overline{K_p}$	$\overline{T_i}$	N	$\overline{K_p}$	$\overline{T_i}$	N	$\overline{K_p}$	$\overline{T_i}$	N	$\overline{K_p}$	$\overline{T_i}$	N	$\overline{K_p}$	$\overline{T_i}$	N
0.1							3.97	7.54		∗	∗		∗	∗	
0.2							2.44	3.22	0.5	2.7	3.0	2.0	2.9	3.13	20.0
0.3				1.97	2.97		2.01	2.39	1.0	2.17	2.34	3.0	2.46	2.37	20.0
0.4				1.64	2.34		1.66	1.9	1.0	1.87	2.0	4.0	2.07	2.05	20.0
0.5				1.41	2.01		1.55	1.65	2.0	1.68	1.69	4.0	1.78	1.79	12.0
0.6				1.15	1.38	0.5	1.36	1.43	2.0	1.45	1.59	5.0	1.48	1.6	6.0
0.7				1.06	1.28	0.5	1.24	1.4	2.0	1.35	1.51	5.0	∗	∗	∗
0.8	1.0	1.9		0.96	1.18	0.5	1.14	1.31	2.0	1.25	1.47	7.0	1.26	1.48	8.0
0.9	0.94	1.36		0.95	1.12	1.0	1.08	1.28	3.0	∗	∗	∗	∗	∗	∗
1.0	0.9	1.35		0.91	1.07	1.0	0.98	1.2	2.0	1.05	1.29	4.0	1.1	1.37	12.0
1.1	0.83	1.23		0.85	1.05	1.0	0.99	1.22	4.0	∗	∗	∗	∗	∗	∗
1.2	0.79	1.22		0.81	1.0	1.0	0.89	1.09	2.0	∗	∗	∗	∗	∗	∗
1.3	0.74	1.13		0.77	0.96	1.0	0.87	1.1	3.0	∗	∗	∗	∗	∗	∗
1.4	0.72	1.13		0.74	0.95	1.0	0.84	1.06	3.0	∗	∗	∗	∗	∗	∗
1.5	0.65	0.87	0.5	0.71	0.91	1.0	0.83	1.07	5.0	∗	∗	∗	∗	∗	∗
1.6	0.62	0.84	0.5	0.68	0.88	1.0	0.8	1.04	5.0	∗	∗	∗	∗	∗	∗
1.7	0.61	0.81	0.5	0.67	0.86	1.0	0.78	1.01	5.0	∗	∗	∗	∗	∗	∗
1.8	0.6	0.83	0.5	0.65	0.83	1.0	0.71	0.91	2.0	∗	∗	∗	∗	∗	∗
1.9	0.58	0.8	0.5	0.63	0.84	1.0	0.69	0.88	2.0	∗	∗	∗	∗	∗	∗
2.0	0.57	0.78	0.5	0.62	0.82	1.0	0.71	0.93	4.0	∗	∗	∗	∗	∗	∗
2.1	0.56	0.76	0.5	0.66	0.84	2.0	0.7	0.9	4.0	∗	∗	∗	∗	∗	∗
2.2	0.55	0.75	0.5	0.64	0.86	2.0	0.67	0.87	3.0	∗	∗	∗	∗	∗	∗
2.3	0.54	0.73	0.5	0.63	0.84	2.0	0.67	0.86	4.0	∗	∗	∗	∗	∗	∗
2.4	0.53	0.72	0.5	0.62	0.82	2.0	0.65	0.84	3.0	∗	∗	∗	∗	∗	∗
2.5	0.52	0.7	0.5	0.61	0.81	2.0	0.64	0.83	3.0	∗	∗	∗	∗	∗	∗
2.6	0.51	0.69	0.5	0.6	0.8	2.0	0.62	0.81	3.0	0.67	0.87	10.0	∗	∗	∗
2.7	0.5	0.67	0.5	0.6	0.78	2.0	0.62	0.8	3.0	0.66	0.86	10.0	∗	∗	∗
2.8	0.5	0.67	0.5	0.59	0.77	2.0	0.61	0.79	3.0	0.65	0.85	10.0	∗	∗	∗
2.9	0.49	0.66	0.5	0.58	0.76	2.0	0.6	0.77	3.0	0.64	0.83	10.0	∗	∗	∗
3.0	0.49	0.65	0.5	0.58	0.75	2.0	0.59	0.76	3.0	0.63	0.83	10.0	∗	∗	∗

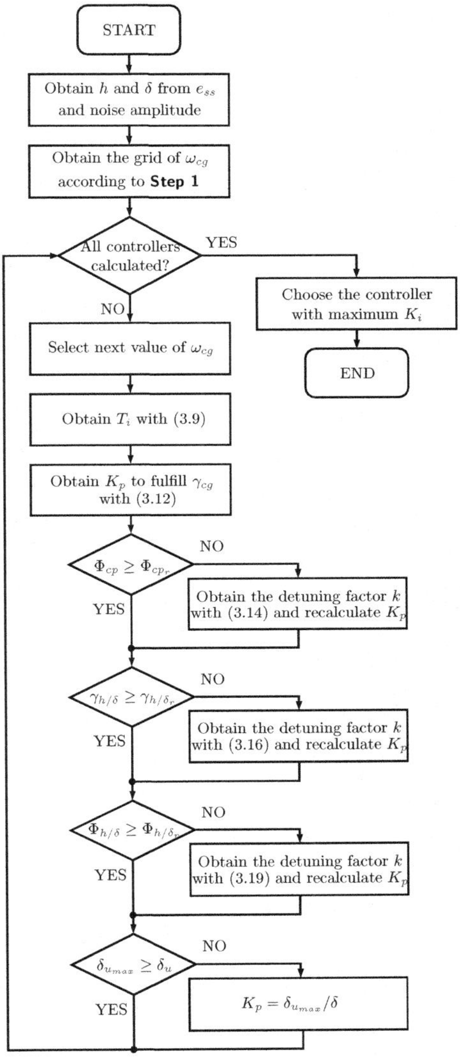

Fig. C.1 Flowchart that summarizes the tuning procedure. (Reprinted from Journal of the Franklin Institute, Vol 358, No. 9, Miguel-Escrig, O., & Romero-Pérez, J. A., *Tuning procedure for event-based PI controllers under regular quantization with hysteresis*, 4643–4670, Copyright (2021), with permission from Elsevier)

Appendix D
Supplementary Calculus

D.1 Classical Specification Dimensionless Calculations

Considering a SISO closed-loop with input disturbance, the relation between the dimensionless error $\bar{E}(\bar{s})$ and the load disturbance $\bar{P}(\bar{s})$ is:

$$\frac{\bar{E}(\bar{s})}{\bar{P}(\bar{s})} = \frac{\bar{G}(\bar{s})}{1 + \bar{G}(\bar{s})\bar{C}(\bar{s})},$$

substituting with the given expressions for $\bar{G}(\bar{s})$ and $\bar{C}(\bar{s})$:

$$\frac{\bar{E}(\bar{s})}{\bar{P}(\bar{s})} = \frac{e^{-\bar{s}}}{\left[\frac{\tau}{L}\bar{s} + 1 + \phi_1\left(1 + \frac{1}{\phi_2\bar{s}} + \frac{N\phi_3\bar{s}}{N + \phi_3\bar{s}}\right)e^{-\bar{s}}\right]},$$

similarly, for a system with dimensions:

$$\frac{E(s)}{P(s)} = \frac{Ke^{-sL}}{\left[\tau s + 1 + \phi_1\left(1 + \frac{1}{L\phi_2 s} + \frac{NL\phi_3 s}{N + L\phi_3 s}\right)e^{-sL}\right]}.$$

Then, the relationship between those two expressions is:

$$\frac{E(s)}{P(s)} = K\frac{\bar{E}(\bar{s} = sL)}{\bar{P}(\bar{s} = sL)}.$$

From this expression it can be obtained the effect of a step disturbance of magnitude 1 on $P(s)$ from a dimensionless step of magnitude 1 on $\bar{P}(\bar{s})$:

O. Miguel-Escrig, *Event-Based PID Controllers with Fixed Threshold Sampling Strategies*, Springer Theses, https://doi.org/10.1007/978-3-031-62505-3

$$E(s) = K\frac{\bar{E}(sL)}{\bar{P}(sL)}P(s) = K\frac{\bar{E}(sL)}{\frac{1}{sL}}\frac{1}{s} = KL\bar{E}(sL).$$

Applying over this expression the linearity property of the Laplace transform, the gain K can be extracted. Then, applying the change of scale of the Laplace transform for dealing with the delay:

$$\mathscr{L}^{-1}\{E(s)\} = K\bar{e}(t/L), \tag{D.1}$$

where

$$\bar{e}(t) := \mathscr{L}^{-1}\left\{\bar{E}(\bar{s})\right\}$$

Being $\overline{IAE}$ the IAE obtained for the dimensionless model:

$$\overline{IAE} := \int_0^\infty |\bar{e}(t)|dt.$$

Using the expression (D.1) to calculate the index for a generic system:

$$IAE = \int_0^\infty |K\bar{e}(t/L)|dt.$$

Extracting the gain from the integral and performing a variable change $t' = t/L$ for solving the integral:

$$IAE = |K|L\overline{IAE}.$$

D.2 IE Validity in Presence of Steady-State Error

$EG\text{-}C(s)$ control loops are characterized by having a static position error bounded by the switching thresholds of the chosen sampling strategy. Thus, the choice of classical IE index to select a controller cannot be used since it tends to infinite. For this reason, a modification of this index has been considered using the sampled error signal $\bar{e}(t)$ instead of the error signal $e(t)$, because it tends to 0 in steady-state, producing the sampled integral error index $I\bar{E}$, defined as:

$$I\bar{E} = \int_0^\infty \bar{e}(t)dt.$$

The control action for this kind of non-linear systems is:

$$u(t) = K_p\bar{e}(t) + K_i \int_0^t \bar{e}(\tau)d\tau - K_d\frac{dy}{dt},$$

which for a PI controller:

$$u(t) = K_p\bar{e}(t) + K_i \int_0^t \bar{e}(\tau)d\tau.$$

Considering a step change at the disturbance input, in steady-state (assuming stability):

$$u(\infty) = K_p\bar{e}(\infty) + K_i \int_0^\infty \bar{e}(t)dt,$$

as $\bar{e}(\infty) = 0$ and taking into account the expression of $I\bar{E}$:

$$u(\infty) = K_i \int_0^\infty \bar{e}(t)dt = K_i I\bar{E}.$$

From the schema block in Fig. 2.1:

$$(u(\infty) + p(\infty))G(0) = y(\infty) = -e(\infty),$$

where $e(\infty)$ is bounded by the steady-state error e_{ss}, and, as a unitary step load disturbance is assumed $p(\infty) = 1$:

$$(u(\infty) + 1)G(0) \leq e_{ss},$$

substituting with the expression of $u(\infty)$:

$$\left(K_i I\bar{E} + 1\right) G(0) \leq e_{ss},$$

taking into account that the IE caused by a load disturbance is $-1/K_i$:

$$I\bar{E} \leq \frac{e_{ss}}{K_i G(0)} + IE,$$

from which the difference between both indexes can be found:

$$I\bar{E} - IE \leq \frac{e_{ss}}{K_i G(0)}.$$

In this last expression, it appears the relation between $I\bar{E}$ and IE indexes. Here, it can be seen that the effect of maximizing K_i minimizes the IE index, and makes the difference $I\bar{E} - IE$ minimum. In addition, as the system is assumed stable, the static error is bounded between:

$$e_{ss} \in \left(-\frac{\delta + h}{2}, \frac{\delta + h}{2}\right),$$

thus, choosing proper values of δ and h, and maximizing K_i, the contribution of e_{ss}, i.e. the effect of the sampling, can be small enough to consider $IE \approx I\bar{E}$, and thus, it can considered the IE as an appropriate selection index for PI tuning.

Appendix E
IEC 61499 Device Models

As stated in Chap. 8, several additional FBs to enable the communication between devices are needed. These communication FBs are implemented in the Device Model of each device.

In Fig. E.1, the device model of the computer's interface is presented. Auxiliary FB to define the interface (FB_LABEL) are presented together with type conversion blocks (REAL2REAL). The communication is carried out by the CLIENT and SERVER FBs, which transmit the data between devices. The set point value is transmitted to the sensor unit with the CLIENT FB and the control action is received from the SERVER FB.

Similarly to the interface case, the device model of the actuator unit is presented in Fig. E.2. In this case, the error value is received from the sensor unit through the SERVER FB, and the CLIENT FB sends the control action to the interface, closing the SERVER-CLIENT pair.

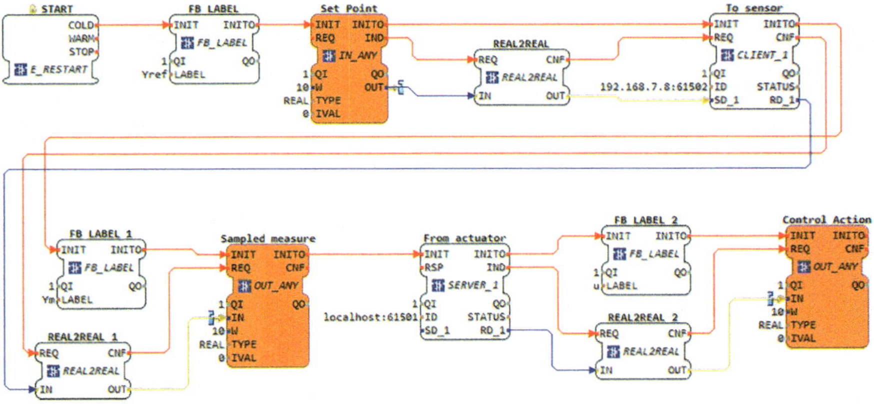

Fig. E.1 Device model of the computer's interface

O. Miguel-Escrig, *Event-Based PID Controllers with Fixed Threshold Sampling Strategies*, Springer Theses, https://doi.org/10.1007/978-3-031-62505-3

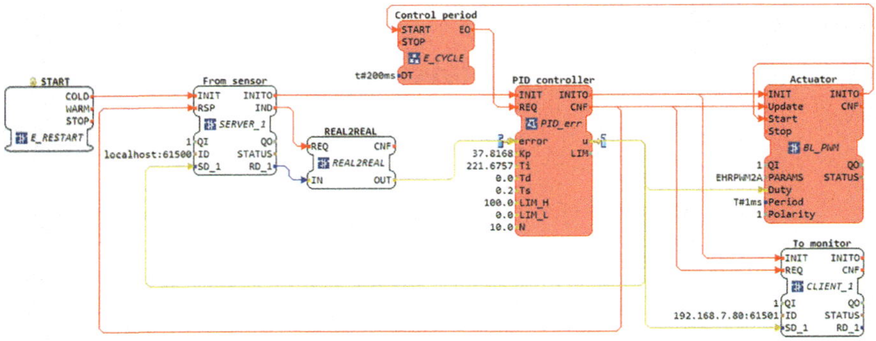

Fig. E.2 Device model of the actuator unit

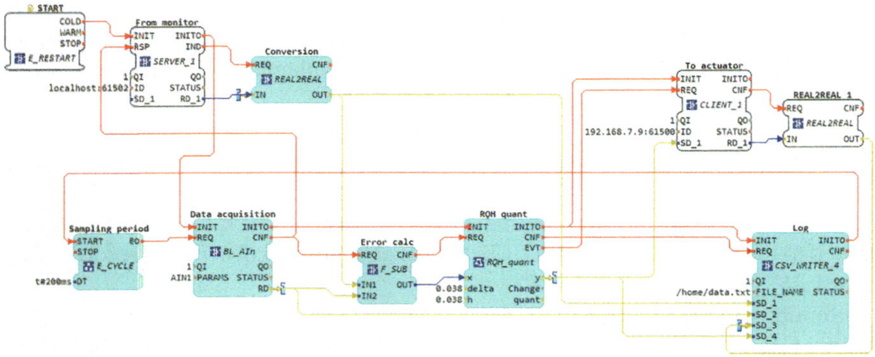

Fig. E.3 Device model of the sensor unit

In Fig. E.3, the device model of the sensor unit is presented. The remaining CLIENT-SERVER pairs are completed. A SERVER FB receives the set-point changes from the interface, and the error signal is sent to the actuator unit with a CLIENT FB. It is important to notice that the error measure is sent to the actuator only when a change in the error signal is detected (EVT event in RQH_quant FB), reducing the data traffic between sensor and actuator units.

About the Author

Oscar Miguel-Escrig is an Assistant Professor in the Department of Industrial Systems Engineering and Design at Universitat Jaume I (UJI) de Castelló. He received the B.S. and M.S. degrees in industrial engineering from the UJI, in 2015 and 2017, respectively, the M.S. degree in automatics and electronics engineering from the Institut National des Sciences Appliquées de Toulouse (INSA), in 2017, and the Ph.D. degree in industrial technologies and materials from the UJI, in 2021. He has been actively involved in research focusing on Networked-Control Systems, Process Identification, Tuning and Auto-Tuning methods, and Automation in IEC 61499 standard.

Throughout his academic career, Oscar Miguel-Escrig has made numerous contributions to the field of Networked-Control Systems, specifically related to the Event-Based Control. His works with regard to this topic have been published in reputable journals and presented at international conferences. He continues working on these research lines collaborating on several research projects.

O. Miguel-Escrig, *Event-Based PID Controllers with Fixed Threshold Sampling Strategies*, Springer Theses, https://doi.org/10.1007/978-3-031-62505-3

Besides, he is interested in the Automation and Control Systems Education, particularly on relevant topics in the current teaching paradigm, such as the involvement of women in STEM studies, incorporating the gender perspective in higher education and artificial intelligence impact and role in the current teaching context.

Oscar Miguel-Escrig looks forward to continuing his research and contributing to the ongoing discourse in the aforementioned interest topics. He can be contacted through his institutional mail for inquiries or collaborations.

SPRINGER NATURE

GPSR Compliance

The European Union's (EU) General Product Safety Regulation (GPSR) is a set of rules that requires consumer products to be safe and our obligations to ensure this.

If you have any concerns about our products, you can contact us on ProductSafety@springernature.com

In case Publisher is established outside the EU, the EU authorized representative is:

Springer Nature Customer Service Center GmbH
Europaplatz 3
69115 Heidelberg, Germany

The manufacturer's authorised representative in the EU is Springer Nature Customer Service Centre GmbH, Europaplatz 3, 69115 Heidelberg, Germany. If you have any concerns regarding our products, please contact ProductSafety@springernature.com

Printed and bound by CPI Group (UK) Ltd, Croydon, CR0 4YY
30/09/2026
02225604-0014